EXPLAINING HUMAN DIVERSITY

Why are humans so different from each other and what makes the human species so different from all other living organisms? This introductory book provides a concise and accessible account of human diversity, of its causes and the ways in which anthropologists go about trying to make sense of it. Carles Salazar offers students a thoroughly integrated view by bringing together biological and sociocultural anthropology and including perspectives from evolutionary biology and psychology.

Carles Salazar is Professor of Social Anthropology at the University of Lleida, Spain. He has a PhD in Social Anthropology from the University of Cambridge, UK.

EXPLAINING HUMAN DIVERSITY

Cultures, Minds, Evolution

Carles Salazar

LONDON AND NEW YORK

First published 2019
by Routledge
2 Park Square, Milton Park, Abingdon, Oxon OX14 4RN

and by Routledge
711 Third Avenue, New York, NY 10017

Routledge is an imprint of the Taylor & Francis Group, an informa business

© 2019 Carles Salazar

The right of Carles Salazar to be identified as author of this work has been asserted by him in accordance with sections 77 and 78 of the Copyright, Designs and Patents Act 1988.

All rights reserved. No part of this book may be reprinted or reproduced or utilised in any form or by any electronic, mechanical, or other means, now known or hereafter invented, including photocopying and recording, or in any information storage or retrieval system, without permission in writing from the publishers.

Trademark notice: Product or corporate names may be trademarks or registered trademarks, and are used only for identification and explanation without intent to infringe.

British Library Cataloguing-in-Publication Data
A catalogue record for this book is available from the British Library

Library of Congress Cataloging-in-Publication Data
A catalog record has been requested for this book

ISBN: 978-0-8153-5652-3 (hbk)
ISBN: 978-0-8153-5654-7 (pbk)
ISBN: 978-1-351-12798-1 (ebk)

Typeset in Bembo Std
by Sunrise Setting Ltd, Brixham, UK

Printed and bound by CPI Group (UK) Ltd, Croydon, CR0 4YY

CONTENTS

List of figures   *vii*
Acknowledgements   *viii*

    Introduction   1

1   Being human   6
    Human evolution in the history of life   6
    Genetic knowledge and individual knowledge   15
    The growth of the human brain   22
    Theory of mind and the modular structure of the human mind-brain   26
    Bibliographical note on Chapter 1   34
    References   35

2   A new form of knowledge   37
    Culture in human evolution   37
    The transmission of cultural knowledge   48
    The problem of meaning   53
    Bibliographical note on Chapter 2   62
    References   63

3   Theories of difference   64
    Understanding human diversity   64
    Opening approaches: anthropology as history   68
    Anthropology against history: functionalism   77
    Structuralism and the problem of meaning again   84

Bibliographical note on Chapter 3   91
References   93

4   Cultural evolution   94
    The puzzle of cultural change   94
    Cultural group selection   101
    Memes and cultural groups   108
    Culture-gene co-evolution   112
    Random processes and purposeful action   115
    Bibliographical note on Chapter 4   119
    References   119

5   Summary and conclusions   121

Index   *127*

FIGURES

1.1	The tree of life	11
1.2	Primate evolution	21
1.3	Human evolution	21
1.4	Growth of the human brain	23
3.1	Evolution of human societies	70
3.2	The race of history	70

ACKNOWLEDGEMENTS

This work was partially supported by the Ministry of Economy and Competitiveness of Spain (grant: HAR2013-48901-C6-2-R). Susan Frekko, Max Holland, Ann MacLarnon and three anonymous reviewers read and commented on earlier versions of the manuscript. I greatly appreciate their manifold witty suggestions and intelligent criticisms. I also owe a big debt to the psychology students of the University of Lleida, for whom this book was originally intended, and to the thousands of questions that they have asked me over the years that made me thoroughly rethink my job as an anthropologist and as an educator. Lastly, my warmest thanks of all go to my wife Núria and my daughter Cecília, without whom nothing of what follows could have ever been written.

INTRODUCTION

In the last few decades, human diversity has become a burning issue in the majority of Western societies. Very different peoples, physically, linguistically and culturally different, have become acquainted with each other for the first time in their history due to massive immigration into Western societies from all over the world. This is a highly remarkable phenomenon that is characteristic of the globalised world in which we live. However, human diversity is nothing new, what is new is the incessant interaction of so many different peoples on a massive scale. In fact, humans have been different from each other practically since the very beginning of human evolution. Anthropology is the social science that studies human diversity. The aim of anthropology is to understand the most distinctive characteristic of the human species: the fact that we humans are very different from each other and, by being so different from each other, that we are also very different from all other species of living organisms.

This book originated in an introductory course to sociocultural anthropology that I taught to psychology students at the University of Lleida between 2010 and 2016. It is 'an invitation' to anthropology rather than 'an introduction' because my purpose in those lectures was not to introduce anthropology to students who might end up studying for a degree in that discipline, but to persuade non-anthropologists of the virtues of the discipline of anthropology. In other words, 'to invite' them to know something about the subject that might prove very useful to psychology undergraduates in particular or to anyone interested in understanding human behaviour. And this is also the purpose of this book: to persuade the reader that human diversity matters and that the social science that deals with this subject is therefore an interesting science.

This accounts for the style in which the book has been written. This is a text for students of the sciences of human behaviour. But it may also be of interest to the general reader who wants to know why humans are so different from each other and what makes the human species so different from all other living organisms. I have tried to develop the main ideas of the book in the simplest and clearest way I could think of, even though the argument will gradually increase in complexity as the book proceeds. There are neither footnotes nor bibliographical references in the main text. There is only a bibliographical note at the end of each chapter. This note is not meant to be comprehensive in any way but is merely a brief reference to some works that I consider useful, interesting or illustrative of what has been dealt with in that chapter.

But this is not perhaps the most innovative aspect of this work. There are lots of other books with a similar purpose in many other disciplines: to invite the reader to see how useful, interesting and amusing that particular discipline is, even or especially, for those who have never thought of taking up a degree in it or studying it in a systematic way. What I take to be its most daring innovation comes with the specific approach I shall try to develop in this book: a biocultural approach. What does that mean?

Human diversity can be accounted for in many different ways. Humans are different from each other, first of all as individuals. These individual differences can in turn be biological: with the exception of identical twins, we are all genetically unique. But most of them are what we might call existential differences. Our life histories are all different, and the experiences and memories we accumulate as a result are also different. Perhaps we do not need any particular science to account for diversity among individuals since most of the time it does not cause any problem. And when it does, that is, when individual differences become pathological in any way, several scientific and humanistic disciplines can deal with them, be they biomedical disciplines when it is a biological pathology or, more generally, psychological or psychotherapeutic disciplines when the individual dysfunction has its origins in an existential problem. In fact it could be argued that all the human sciences and the humanities, philosophy included, are there in one way or another to help us address these existential problems.

But far more problematical and upsetting than individual differences are collective differences, namely, those that make whole groups of individuals different from other groups. This has been the traditional object of the analysis of anthropology. Admittedly, the difference between individual and collective differences may not be as clear-cut. But there is always the sense that collective or generalised phenomena are less likely to be the product of contingent events than individual particularisms. And the same applies to the distinction between individual and collective differences. Thus, insofar as those collective differences can in turn originate in two relatively independent sets of causes, biological or cultural, two separate sub-disciplines within anthropology take those two discrete sets of collective differences as their object of study. Biological or physical anthropology deals with biological collective differences within the human species, and social, cultural or sociocultural anthropology (they are all different names for the same discipline) studies the causes of human diversity when they have, or are thought to have, a cultural origin. As we will see in what follows, intraspecific biological differences (i.e. biological differences between the members of the same species) are not particularly distinctive of the human species. In general terms, we humans are not biologically more different from other humans than the member of any other species of living organisms is from any other member of the same species. In fact, the very opposite seems to be the case: humans are more genetically homogeneous than other primate species. Biological anthropology deals with humans as a single species, and in studying what might be intraspecific differences, biologists are seeking unifying underlying principles, relying on comparative analyses with other primate, mammalian and vertebrate species. Relative intraspecific genetic (and biological by implication) homogeneity in humans could be the reason why most biological anthropologists study human differences across time rather than across space. For it is when we go back in the history of the human species that human biological diversity starts to become interesting. Because, as is the case for any other species of living organism, it is from those ancestral differences that the current biological characteristics of human beings can be accounted for.

Cultural diversity, by contrast, even though it also exists in other species such as chimpanzees, is much greater in humans and it makes the members of the human species more different from each other than the member of any other species from any of their conspecifics. Now the question is what kind of relationship can be established between human biological diversity and cultural diversity. Once anthropologists reached the conclusion (and rightly so) that cultural differences cannot be accounted for by biological differences, for there is not enough biological diversity among human populations to explain the vast diversity seen in cultural behaviours, cultural anthropology started to develop as a totally independent discipline from its sister discipline, biological anthropology and from any other discipline within the biological sciences. And this is when things started to go awry. It was not only cultural anthropology that set itself apart from the biological sciences. This, in fact, has been the general mood among the social sciences and the humanities whose roots can be traced much further back. The idea that anything having to do with human behaviour has nothing to do with human biology originates in a totally anti-scientific conception of human beings as being radically, absolutely, different from any other living thing, as if they had been made of a different substance. This is clearly not true. As we will see throughout this book, humans are in many ways unique and different from the other species of living organisms, but humans are biological beings through and through, and the origin of that uniqueness can be fully accounted for in biological terms. This does not mean that cultural differences are caused by biological differences. We have already seen that that is not the case. But let me emphasise this point: even though there are some cultural differences that might have some underlying biological cause, such as the relationship between lactose tolerance and pastoralism (more on this in Chapter 4), that is not the case for the overwhelming majority of them. This does not mean, however, that those cultural differences do not have their origins in the *general* biological characteristics of the human species – not in the specific biological characteristics of particular groups of human beings. And this is what, in my view, a biocultural approach to anthropology and to human diversity should throw into relief.

This infamous ignorance of the biological foundations of human behaviour by the social sciences in general, which became ignorance of the biological causes of human cultural diversity by cultural anthropologists in particular, was perhaps not terribly significant 30 or 40 years ago, when the biological sciences of human behaviour were still at an early stage of development. At that time, we could produce strictly cultural, that is, non-biological, explanations of human behaviour because little was known about the underlying biological causes of that behaviour. But the situation is now radically different. Since, approximately, the last two decades of the twentieth century, the study of human biology has undergone immense development. We should include within human biology, among others, the disciplines of genetics, evolutionary biology, evolutionary psychology and the neurosciences. And this enormous development has gone unabated, even accelerated, in the twenty-first century. So, ignorance of this huge amount of knowledge produced by those disciplines can no longer be justified. To make matters worse, parallel to those developments in biology, especially also in the last two decades of the twentieth century, a growing anti-scientific mood started to take shape in the social sciences and the humanities, mainly, I would even say, among cultural anthropologists. This was an infamous consequence of the so-called postmodernist critique, which has done so much harm to the social sciences and the humanities of the late twentieth and early twenty-first centuries – in addition to, it must be conceded, a few good ideas that came from that intellectual movement, but I cannot dwell on that now. Some cultural anthropologists (not all of them by any means) not only continued to be ignorant of the staggering developments that were taking place in the biological sciences, but they even felt proud to be so, as if that kind of knowledge was a sort of evil knowledge that had to be banned or proscribed in any way for political or moral reasons, very dubious political and moral reasons indeed.

That was a totally aberrant situation whose effects are still apparent in the present-day intellectual climate of the social sciences and the humanities. This book tries to be a modest contribution to overturn that dreadful development. Very few introductory texts to anthropology take an integrative biocultural perspective in which both the biological and cultural dimensions of human diversity and human behaviour are given their due. And this is precisely the purpose of this book. My aim is to provide a brief, simple, jargon-free and concise account of human cultural diversity, of its causes and the ways in which anthropologists go about trying to make sense of it. But this is going to be an account of human diversity firmly rooted in the biological characteristics of human beings.

The book is divided into five chapters. The first one, 'Being human', presents a succinct analysis of the main biological characteristics of the human species in comparison with other species of living organisms set within an overall evolutionary framework. Basic principles of evolutionary biology are also explained in this part in very plain, simple language accessible to non-specialists. The evolution of the human brain constitutes one of the central themes of this first chapter, for the brain of human beings is the biological characteristic of the human species that makes it very different in quantitative terms from all the others. In the second chapter, 'A new form of knowledge', we shall explore what could be defined as the qualitative difference between the human species and the rest. And this is cultural knowledge, the special form of knowledge that the big brain of human beings is capable of producing. Culture is taken here as an emergent property of the human brain, that is, something that originates in the specific characteristics of a biological organ but that at the same time transcends those characteristics. Culture introduces us into the intriguing notion of meaning, the constitutive factor of culturally determined human behaviour. In the third chapter, 'Theories of difference', we shall be looking at how anthropologists go about studying cultural meanings and what theories they have produced to explain those cultural meanings. Only the so-called classical theories in social and cultural anthropology will be dealt with in this chapter. My purpose is to identify broad theoretical arguments in each of those schools of thought that can pave the way for the integrative biocultural perspective espoused by this book. Finally, the fourth chapter, 'Cultural evolution', deals with one of the most controversial issues in the study of human behaviour and human cultural diversity. Human cultures differ not only across space but also across time. But what relationship can be established between this synchronic diversity and diachronic diversity? Is cultural history a totally contingent process? Or is there something akin to natural selection in biological evolution that explains why certain cultural forms seem to be more successful than others? This is perhaps the most theoretically dense and intellectually demanding part of the book. But hopefully at this stage the reader will have gathered enough conceptual tools to enable them to come to grips with surely one of the hottest research topics in current social sciences. The book concludes in the fifth chapter with a brief summary of the fundamental ideas developed throughout the text.

1
BEING HUMAN

Human evolution in the history of life

Imagine a Martian scientist with the job of studying all life forms on planet Earth. In all probability, the first thing she would do would be to classify living organisms according to the species to which they belong. But when she came across the human species, something very odd would immediately catch her eye. The human species seems to be very different indeed from all the others.

In all likelihood, this imaginary alien scientist would not be much impressed by the apparent results of human intelligence – whatever we mean by that and whatever the Martian understood by that. True, some humans are capable of making very spectacular things, such as big buildings. But termites and corals accomplish similar feats and with infinitely smaller brains, or with no brains at all. Furthermore, imagine that the Martian could have – by some sophisticated technological means unknown to us – the whole span of human history in front of her. She would then see that the majority of humans have lived for much of human history in quite humble dwellings, not dissimilar to birds' nests. Maybe it is the characteristics of the human body that the Martian would find odd. For instance, humans have very big brains. But it so happens that a whale's brain is seven times bigger than a human brain. It is not even the relationship between the brain weight and the weight of the rest of the body, since in the case of the shrew, the brain weight corresponds to ten per cent of the body weight, whereas for humans it is two per cent. We shall come back to this.

What about culture? That is certainly a good point, as we shall see in more detail below. But the question is: how would the Martian know what 'culture' is? Why should we define human buildings as 'culture' but not a termite nest, or a beaver's dam or coral reefs? How would the Martian see what is culture and what is not? Without a doubt, before coming to terms with this somewhat obscure concept (and the same applies to the concept of intelligence that we have just mentioned), there is something much simpler and obvious that would draw the Martian's attention, namely, human diversity. In no other species would she find the diversity characteristic of humans – again, above all if she was able to observe the whole of human history on the spot. That is diversity of ways of life. Any animal of any species leads a way of life practically identical to the way of life of any other member of its own species (setting aside now domesticated animals). The life of a gorilla, ways of feeding, interacting with their environment, with other gorillas or other animals, ways of mating, reproducing, etc. is practically identical with the life of any other gorilla. And the same applies to any other animal. Not only that, the way of life of any member of any species has been practically identical to the way of life of any other member of that same species as far back as the history of the species goes. A gorilla or a chimpanzee has the same way of life now as they had 1,000 years ago, or 100,000 years ago – or at least we have no evidence to suggest that it was not.

The situation with humans is totally different. We not only find diversity in ways of life at any time in history, but if we go back in history we realise that humans' ways of life have been changing all along. The life of any human being now is very different from the life of his/her ancestors 100 years ago, never mind 1,000 years ago or more. The question is why. What could account for this diversity in the human species? Why are we humans so different from each other as compared with other animals, even with animals genetically very close to us, such as other primates? How do we explain this diversity, which clearly appears as the most distinctive characteristic of the human species?

Now we come to the concept of culture. Anthropologists believe that what makes humans so different from each other – and by making them so different from each other it makes them even more different from all the other species – is something we call 'culture'. Note that 'culture' is a concept. It is not something that we see, it is just an idea. It is something we cannot see and that we use in order to understand something that we can see: human diversity. But there is more than one kind of human diversity. A sub-discipline within anthropology is what is known as physical or biological anthropology, which consists of studying human diversity from a strictly physical or biological point of view. Humans are also different from each other in biological terms. But these differences are no bigger than the same differences that we can find among the members of any other species. In fact, humans are more genetically homogeneous than most other mammalian species. In other words, intra-specific biological differences, though they exist, are not a distinctive characteristic of humans.

At one time, it was thought that human diversity could be explained in biological terms, that humans who had different ways of life were biologically different. This was so because, apparently, humans who have very different ways of life from Westerners, the so-called 'primitive peoples', happen to be also physically different, they belong to what used to be called a different 'race'. A race is just a group of individuals, belonging to the same species, who share some, visible, physical characteristics (particularly skin colour, shape of eyes, of the head, body, etc.) believed to be inherited, that is, physical characteristics that one gets from his or her parents. Now, it has been widely demonstrated that the so-called racial differences among humans do not explain human diversity of ways of life. What is wrong is not that human differences have nothing to do with human biology, which they do, as we will see, what is wrong is that *human differences* or human diversity of ways of life is the result of human *biological* differences.

So how do we explain human diversity? If it is not biology, what is it? Anthropologists believe, and have believed this for a very long time, that human diversity, which is so distinctive and unique to the human species, can only be explained by means of a concept that refers to something also distinctive and unique to the human species, and that is the concept of culture. Anthropology is sometimes also called 'the science of culture', and anthropological theories are called 'culture theories', even though by rights it should be the science of cultural difference rather than the science of culture. A way of talking about the objectives of this book would be to explore the concept of culture, that is, to understand how culture intervenes in the determination of human behaviour. Now, in the determination of human behaviour many factors intervene (biological, biographical, existential, etc.). Among all these factors there is one we call 'culture', the culture factor. And the purpose of this invitation to anthropology is to understand how this works, how the cultural factor intervenes in the determination of human behaviour.

Let us put forward a provisional definition of culture: a system of meanings and symbols in terms of which humans rule their behaviour. Whenever and wherever we see someone behaving in such a way that we can explain that behaviour in terms of a particular system of meanings and symbols, we will call that behaviour 'cultural behaviour', we will say that that behaviour has been 'culturally determined'. Now where does this system of meanings and symbols come from? Why do humans have cultures and other animals do not? Or do they have cultures? How do cultures enter into our heads? How do we 'understand' a culture? These are some of the questions we will be looking at in the following pages. As we shall see, cultures have many distinctive characteristics. But there is one of them I would like to underscore right from the beginning: those systems of meanings and symbols that compose human cultures are not innate, that is, they get into human minds as a result of a process of learning. We learn our cultures, we are not born with them, and we learn them as we grow up in a particular society. That is what makes culture different from biology, we are born with our biology (setting aside for now postnatal environmental differences), we are not born with our culture. Since human societies can be very different, and have been very different throughout human history, the cultures that humans learn as members of those societies are also different. Consequently, the behaviours that result from these different cultures are also different. So this is, in a nutshell, the way we anthropologists normally explain human diversity. Humans are different, so different, from each other because the cultures they learn are different.

What we will try to find out now is what makes the human brain susceptible to produce and to learn, to absorb a 'culture'. Why is it that humans have cultures (and hence they are so different from each other) and other animals do not – or if they do, it is always to a much more limited extent than humans? What happened in the process of human evolution that made human brains capable of producing cultures and learning, that is, assimilating, those cultures?

Replicating structures

Let us start the story of human evolution at the very beginning. The process that gave rise to the appearance of the human species is called the process of human evolution. The point I would like to emphasise here is that it was during this long process, which lasted approximately six million years (we will see in a minute why we say that it lasted six million years), that culture came into existence, for the first time in the history of this planet, and, as far as we know, for the only time. So it was a very unique thing to happen in the 3.5 billion years of history of life on earth.

We say that human evolution lasted for six million years because it is reckoned that the last common ancestor of both humans and our closest living species now (chimpanzees) lived six million years ago. This means that if chimpanzees became extinct, then human evolution would have lasted for ten million years instead of six, for that is the time when we would find our last common ancestor with gorillas, which would be our closest living species in the absence of chimpanzees. Correspondingly, if Australopithecus were still alive, then human evolution would be shortened down to two million years. Now, in whatever way we happen to measure the length of human evolution, what cannot be doubted is that something happened during this process that explains the emergence of cultures, something that had never happened before, and that will never happen again afterwards. It does really look like a miracle. But that is not the only 'miracle' that took place in the history of our planet. Think of the origins of life itself. Life on earth began 3.5 billion years ago. Our planet is approximately, according to astrophysicists, five billion years old. That is, it came into existence 'only' 8.7 billion years after the big bang, which is supposed to be the beginning of the universe. When life began on earth 3.5 billion years ago, pretty much like culture, it did it only once. That explains why all living organisms that exist right now on this planet come from other living organisms, which in turn come from other living organisms and so on and so forth until the very beginning of life, in other words, until the very first living organism that existed on earth, which was not a living organism properly speaking, it was just a sequence of a nucleic acid that started to replicate, that is, to produce copies of itself. This is what goes by the name of LUCA: Last Universal Common Ancestor. We still do not know very well why this thing happened 3.5 billion years ago and it did not happen again after that (and, probably, we will never know for certain). But, again, what we do know is that it was something very unique, that it has happened once in five billion years and has not happened again (or at least it has only happened once on earth).

Both the appearance of life on earth and the appearance of the human species, that is, the process of human evolution with this very idiosyncratic and unique characteristic (which is that it is a cultural species, strictly speaking, not the only cultural species that has ever existed but undoubtedly the most cultured), are both facts so unique and mysterious in themselves, that we still do not have a clear explanation as to why this was so. All this makes creationist explanations look rather attractive, both for the origins of life and for the origins of the human species. Creationism is the doctrine according to which neither the origins of life nor the origins of the human species – nor the origins of any other species for that matter – can be explained without resorting to some sort of external (divine) intervention. Needless to say, this is not the point of view taken in this book. No matter how attractive creationism might be – and it certainly is, given all we have said, given all these mysteries and uncertainties concerning the origins of life and of our species – it is not a scientific hypothesis because it takes as a premise something that cannot be proven: either the existence of supernatural beings or of any other being to whom we can attribute creation powers.

Here we are not concerned with the origins of life as such (not directly concerned, at least) but with the origins of the human species, and with what is so distinctive about the human species, what makes this species so different from all the other species that have existed on this planet. Still, I think it is important to keep all of these temporalities, these huge time dimensions, in mind while we think about the origins of our species. The following panel aims to be a concise summary of the history of life on earth:

What we see in Figure 1.1 is what Darwin called 'the tree of life'. This very simple graph shows that all living organisms, humans included, share a common ancestor and this was that ancestral acid strip that came into existence 3.5 billion years ago. Let us take a very quick look at how this process of the evolution of life took place. This a well-known story so I do not claim any originality for what follows. My interest is just to emphasise a few particular aspects that will be relevant for what we will see later on.

We shall begin with the very concept of life itself. What is a living thing? A living thing is a structure that replicates itself, that is, that is capable of producing copies of itself. Nowadays we know of only one substance that is capable of doing this, it is a molecular chain called DNA. All living organisms contain DNA. Now this DNA is, in turn, made of what we call nucleotides. There are four of these nucleotides: adenine, cytosine, guanine and thymine. In a DNA molecule what we have is a long strip made of a sequence of these nucleotides arranged in a particular order:

ACGTGGTAAGCTAACTGGA, etc.

So this is the substance of life, what life is made of. These DNA strips are the structures that reproduce themselves. What is the purpose of these DNA strips? Why do they have to be ordered in particular sequences? Because, together with other chemical processes that do not concern us now, these particular sequences of DNA allow for the synthesis of another chemical, another substance that constitutes the bricks, as it were, with which the body of any living organism can be built. These

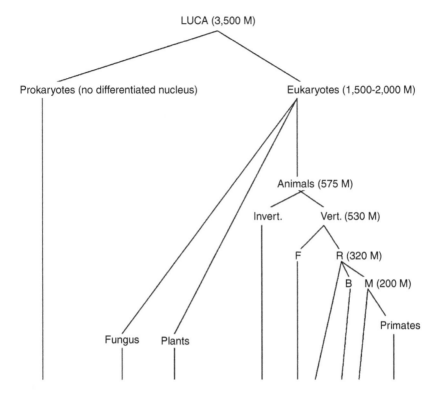

F: fish
R: reptiles
B: birds
M: mammals

FIGURE 1.1 The tree of life

are the proteins. The body of any living thing, ourselves included, is made out of proteins – plus other components such as fats, water, etc. Important differences between one organism and another originate in the type of proteins of which organisms are made. That is why DNA, ordered in specific sequences, is so important for life. DNA is like the alphabet and proteins are the words that can be made with the letters of this alphabet: the letters are always the same, but the words are different, hence the bodies of those living organisms are also different.

Now, what makes life such a unique phenomenon it is not so much the substance it is made of (DNA and proteins) but the power of this substance, the power to produce copies of itself. Remember that a living thing is just a structure that can produce copies of itself. How does this process of reproduction take place? In two different ways. The most simple and earliest way of reproducing a living organism is by cloning, which consists merely of that organism producing a copy of itself, which we call a clone. But approximately 1,000–1,200 million years ago a new form of reproduction came into existence among the eukaryotes (those whose cells have a nucleus, see below). That was sexual reproduction. In sexual reproduction, we need not one organism but two organisms, the progenitors, that will exchange their DNA (half of it) in order to produce a new organism, which will not be a replica, a clone, of any of its progenitors, but a mixture of the two.

Sexual reproduction takes place among eukaryote organisms, and a eukaryote organism is one which is made of cells with a differentiated nucleus. They can be either unicellular organisms, with just one cell or pluricellular organisms, like ourselves (who, on average, have 100 trillion cells or 10^{14}). In the nucleus of each of these cells we have our DNA stored in huge molecules called chromosomes. We humans have 23 pairs of these chromosomes. We will see later why these chromosomes come in pairs in all sexually reproducing animals. Now, when sexual reproduction appeared in the history of life it rapidly became so successful that the majority of eukaryote organisms ended up reproducing themselves in this way (even fungi and plants – though there are still some of these which reproduce by cloning). Sexual reproduction was so successful because by mixing the DNA from two other organisms it created a new one which was different from the first two, that is, with new DNA strips and, consequently, with new combinations of proteins, and, as a result of that, with a body slightly different from that of its progenitors, that is to say, with new possibilities of adaptation to the environment and, specifically, to changing environments.

In order that a living organism may reproduce itself (a DNA strip may produce copies of itself) we need a particular environment. Animals, for instance, need organic matter, which they get from other living organisms, be these other animals or plants, to feed themselves, live and reproduce. Plants and fungi are capable of synthesising inert matter (oxygen, carbon dioxide, mineral salts, etc.) and turn it into organic matter. But to do all that, organisms need a particular environment – they have to find in that environment the materials they need to live. Adaptation is the condition of those organisms that are able to find in their environment what they need in order to live and reproduce. Non-adaptation or maladaptation is the condition of those organisms that cannot find in their environments the substances they need to survive and reproduce.

It is time now to introduce a new concept which will enable us to better understand how the process of life takes place on our planet. This is the concept of natural selection. Now, when living organisms reproduce, either by cloning or by sexual reproduction, DNA makes copies of itself. But sometimes there are mistakes in producing these copies. A mistake takes place when the original sequence was ACGGTAT and the copy is ACGGTGT, so an A is changed into a G. This is what biologists call mutations. A mutation is thus simply a mistake in the replication of a particular DNA sequence. A mistake entails that the new mutant DNA sequence might not be able to produce, to code for, the same protein as the old sequence. The majority of these mutations are either harmless or non-adaptive, which means that the mutant individual is not capable of reproducing itself as well as, or at the same rate as, non-mutant individuals. The result will be that the mutation will disappear, immediately if the mutant cannot reproduce, or after a few generations if the mutant reproduces itself less than non-mutants. But it so happens that on some occasions the mutation might turn out to be adaptive, in which case the mutant individual is capable of reproducing itself more than non-mutant individuals. The result will be the reverse: after a few generations only mutant individuals will exist in that particular environment. They will have outreproduced the others. If this process is repeated millions and millions of times as living organisms keep reproducing and keep producing mutations, the final result will be that all organisms appear to be very well adapted to the environment in which they live, and this is so because non-adaptive mutations are selected against whereas adaptive mutations are selected in. Now then, this does not mean that all individuals end up genetically identical. Copying mistakes or mutations occur even when individuals are already well adapted to their environment, so new variants continue appearing, to be tested against that environment again and again. This is the process that since Darwin, who was the one who discovered it (even though he did not know anything about DNA), goes by the name of natural selection. And the result of this natural selection, which is the progressive adaptation of organisms to their environments, is what is called, also after Darwin, evolution. It should be noted, however, that adaptation is always a slippery concept. Maybe the result of natural selection should be more properly defined as merely the differential reproduction of a population of variants, for the more finely tuned to a specific environment an organism happens to be (the more 'adapted'), the more prone to a disaster when the environment changes. That is why some of the species most likely to evade extinction are not tightly adapted to any particular environment (rats, cockroaches, etc.).

Now, for natural selection to take place, we need to keep changing the DNA structures (and hence the proteins produced by them) of the organisms that need to be adapted at a certain rate, so that the good ones are kept and the bad ones are discarded. Whereas organisms that reproduce by cloning can only change their DNA structures through mutations, organisms that reproduce sexually keep changing their DNA structures at each generational replacement by the very process of sexual reproduction, for new individuals are always a combination, a mixture, of the DNA structures of their two progenitors. In other words, sexual reproduction increases the variability of DNA sequences within a given population, and by so doing permits a better and faster adaptation through the process of natural selection. That is the reason why the majority of living organisms reproduce sexually, because once sexual reproduction appeared in the history of life it spread very rapidly, and it did so because it allowed a better and faster adaptation to sexually reproducing individuals.

Let me now introduce another concept that we have been implicitly using so far but which it is now time to make explicit. That is the concept of gene. For a very long time, it was thought that a gene was a protein-producing DNA sequence. But in the last few decades and, especially, since 2002, which is when the mapping of the human genome was completed, biologists realised that most of our DNA, most of the DNA of any particular organism, is what they call 'junk DNA', which is DNA that does not code for any protein and that, apparently at least, does not seem to fulfil any function other than making copies of itself. In the case of humans, it has been calculated that out of the 3 billion DNA bases in our genome (a genome is the sum of all the DNA sequences of an organism), only one per cent are protein-producing DNA sequences, nine per cent are DNA sequences that do not produce proteins but that help, in one way or another, in the production of those proteins, in other words, they fulfil some kind of function, even though it is not the production of proteins. But the remaining 90 per cent is junk DNA, which means that, for the time being, 90 per cent of our DNA does not seem to fulfil any function at all. British biologist Richard Dawkins compares this junk DNA with the files of a computer. Most of the files we have in a computer are junk files, files that may have been useful once but then we forget about them and just leave them there. So the amount of files, the amount of actual information that we are using from a computer at any one time is just a tiny fraction of all the information that that computer has stored in its hard disk. So our body (our genome) is like the hard disk of a computer, with a lot of information that is totally useless, even though it might have been useful at one time in our evolutionary history, useful to some of our ancestors.

So what is a gene? A gene is a DNA sequence that goes unmodified (except when there is a mutation) from one generation to the next. The reason why it goes unmodified is because it fulfils a function now, such as coding for a protein, or it did in the past. Consequently, a gene could be defined as a unit of heredity, because the same DNA chunk will travel down the generations unmodified, reproduction after reproduction. In the case of humans, for instance, we have 23 pairs of chromosomes in the nucleus of each of our 100 trillion cells in all of them, except in one particular group of cells that we call germ cells. These are sperm cells for men and ova for women (also called 'gametes'). These are different from the other cells (called 'somatic cells') in that in their nuclei they only have 23 chromosomes, not 23 pairs. This is so because when sexual reproduction takes place, male DNA moves into the female nucleus in order to produce a new cell (called a zygote), out of which a new individual will be born in due course. We all have 23 pairs of chromosomes in somatic cells simply because one member of each pair comes from our father and the other from our mother. Now this zygote, which is just a fertilised egg, will start dividing itself up in a process called mitosis thus producing more copies of the initial cell. That is how it will eventually grow into an embryo, a foetus and a new human being once it is born. Now alongside this mitosis, at one particular stage, a new process will start up, and this is the so-called meiosis. Meiosis is different from mitosis in that instead of producing a new cell with 23 pairs of chromosomes it produces a new cell with only 23 chromosomes, in other words, it produces the germ cells of the new individual (sperm cells or ova, as the case may be). Meiosis takes place by means of a process called 'crossing-over', in which whole chunks of the DNA of one particular chromosome get mixed with whole chunks of the DNA of its 'partner' chromosome, so to speak, that is, the other member of the pair that comes from the other progenitor. These whole chunks that do not get broken down during this process of crossing-over are the genes.

Once we know what a gene is, and what DNA, adaptation, natural selection and evolution are, we already know the basics of what life on this planet is all about. We are in a position now to understand a new concept, a very fundamental concept for the argument to be unfolded in much of this book. And that is the concept of genetic knowledge. First of all, I define as 'knowledge' any piece of information – either coming from the environment or information internal to the body of a particular organism – which is somehow processed by that organism and, as a result of that, gives rise to some kind of change in that organism. It can be a behavioural change in the case of animals, or any other change in the structure of that organism, which in any case can be seen as a sort of response to that information. So knowledge is just *information processed* by an organism.

My thesis is that in the case of humans, our behaviour is determined by, or can be explained as being the result of, three different kinds of knowledge: genetic knowledge, individual knowledge and cultural knowledge. These are kinds of information lodged somehow in our organism, information that is processed by this organism and that determines its behaviour. So the behaviour of any human being is the result of the interaction of these three different kinds of knowledge. A key question to be addressed in what follows is how this interaction actually takes place. Let us take a closer look at the first two: genetic and individual knowledge.

Genetic knowledge and individual knowledge

Genetic knowledge is knowledge stored in our genes, in our genome, as a result of the process of natural selection. As we have seen, through natural selection, bad mutations in our DNA sequences are discarded while the good ones replicate themselves by passing down onto the next generation. Thus, at the end of the day, what we have in the genome or any organism, in addition to the junk DNA we have already referred to, are those DNA sequences, those genes, that happen to be adaptive, which means that they code for the set of suitable proteins that allows for the reproduction of that organism in a given environment; and they have allowed for this reproduction for generations and generations. This is the 'genetic knowledge' of that organism, a form of knowledge that determines much of the characteristics of that organism, both of its body and, in the case of animals, of its behaviour. I call it a form of knowledge because it is information from the environment that has been processed by the organism and given rise to a particular response, which is a particular DNA sequence or gene or set of genes, which, in turn, code for a certain biological trait (somatic or behavioural) that turns out to be adaptive in a given environment. Strictly speaking, it is not the environment that produces DNA mutations, for these are totally random. The environment merely selects in the good ones and selects against the bad ones. The important thing to emphasise here is that all these characteristics exist, all this genetic knowledge exists, because it permits the adaptation of this organism to a given environment, that is, its survival and reproduction. That is what genetic knowledge is all about. Because genetic knowledge comes (very often, though not always, as we will see) from hundreds of thousands, even millions, of years of evolution through natural selection, it can be very precise. Consequently, the adaptations that we find in living organisms look almost perfect, as if they had been designed by a master engineer.

There is a branch of Creationism (remember, the doctrine according to which we cannot explain the existence of life on earth without alluding to some sort of divine intervention) that is called 'Intelligent Design'. According to the supporters of this theory, living forms on earth are so complex and appear to be so finely tuned to their environment that we need to postulate the existence of some sort of intelligent designer as their cause. It all started at the beginning of the nineteenth century, when an English theologian by the name of William Paley published a treatise on natural theology where he tried to demonstrate the existence of God in precisely these terms of the need for an intelligent designer. He said that if we find a stone in the middle of the road, we do not naturally think that it has been placed there deliberately by someone, or that it has the shape it has because it has been so designed by someone with some purpose. We think that the stone is the way it is totally by chance. But if we come across a watch, we will not think that this watch with all its pieces have been put together totally by chance. That is impossible. As if we could throw up a pile of scrap and when it falls down, by mere chance, it forms a watch. There is only one possibility among trillions and trillions of trillions of this happening. In other words, to have a watch we need a watchmaker. The same applies to living organisms, Paley thought. The organ of the simplest living thing is also extremely complex, sometimes much more complex than the most sophisticated of our watches. Think of a human eye, for instance, the number of tiny cells, tissues and nerves all finely tuned and articulated to allow us to see. Can this be made 'by chance'? We need a maker for the eye, or for any other part of a living organism. And this, according to Paley, can only be God.

But now we know better. There is no need for any God or intelligent designer for natural selection to take place. The simple fact that natural selection discards the bad mutations and only keeps the good ones means that after a few generations those good mutations can be very good indeed. But the process is automatic, no mastermind is needed for it to take place. Furthermore, living organisms do not need to be conscious in any way of their genetic knowledge. We can say that genetic knowledge is simultaneously very precise and stupid. 'Competence without comprehension': this is the way American philosopher Daniel Dennett has defined the outstanding feat of biological evolution. And it has one single purpose: the replication of DNA sequences.

Four main characteristics distinguish genetic knowledge from other kinds of knowledge (which we will see very soon):

1. It is innate: we are born with our genetic knowledge, we do not have to learn it from anywhere. It could be argued that it was 'learned' by our ancestors and passed down to us in our genome.
2. It is accumulative: it is the result of the accumulated experiences of all the generations that pre-existed us.

3. It is shared. We share most of our genes with all the other members of our species. Genetic variation in the human species is, on average, one per thousand DNA bases. Some of it is also shared with other species, particularly those closely related to us, that is, with whom we have a common ancestor who lived not too far back. We share 98 per cent of our DNA with chimps. In other words, with the exception of this two per cent, which is specific to humans, all the rest is shared with other non-human species: the closer our common ancestor is, the bigger the amount of DNA that will be shared. Notice, incidentally, that it is 98 per cent of human DNA that is shared with chimps, not 98 per cent of human genes.
4. It is automatic: the somatic or behavioural characters that genetic knowledge gives rise to are generated automatically with the appropriate stimulus. For instance, when our genetic knowledge informs us that we can get burnt if we approach a very hot thing, as soon as we feel the heat, our nervous system will automatically push our body away from the source of heat.

Our interest is specifically in that bit of genetic knowledge that determines behaviour. But the point I wish to emphasise is that the mechanism that makes a particular DNA sequence generate a particular behaviour is exactly the same as the one that generates any other characteristic of the organism. The set of behaviours generated by genetic knowledge are called instincts or instinctive behaviour. An instinct is a drive of our nervous system that makes us react or behave in a particular way when we receive a particular stimulus, a particular piece of information, from the environment, for instance, our drive to withdraw immediately the parts of our body that come into contact with something very hot.

The mechanism is the same, and the purpose is also the same: to enhance the adaptability of the organism. That is to say, all our instinctive behaviours exist because they enhance our capacity to survive and reproduce in a given environment. Here is another example: little children are instinctively afraid of the dark. We are all afraid of the dark, in fact, when we enter a dark space (a dark room, etc.) our whole nervous system becomes very alert, automatically, we do not need to have had a bad experience in the dark to be afraid of it. And the same is probably true of other animals, particularly other primates, which, like us, rely basically on their sight to orient themselves – unlike most mammals, which rely mainly on their sense of smell. It is easy to find an explanation: in a dark place, primates are very vulnerable to the attack of predators, so we might suppose that in ancestral times those of our ancestors that were not afraid of darkness or were not alert when it got dark had fewer possibilities of survival, that is, less possibility to pass down their genes.

Notice that for genetic knowledge to emerge, that is, for adaptive characteristics to be generated by our DNA, we need a lot of mutations to take place, so the good ones are kept (selected in) and the bad ones are discarded (selected out). And for a lot of mutations to take place we need a lot of reproductive cycles in a stable environment. So the longer the reproductive cycle of a particular species is, the more stable the environment will have to be (more stable for a longer time) for that species to produce adaptive traits, that is, to produce genetic knowledge. And, conversely, the shorter the reproductive cycle of a particular species (think of micro-organisms that can reproduce themselves in a few minutes after having been born, whereas it might take 20 years or more for a human being to reproduce), the faster its adaptation takes place, that is, the faster it will acquire the characteristics that will enable it to adapt to that particular environment and, consequently, the faster this particular species will adapt to a *changing* environment.

For any species, the environment of evolutionary adaptedness (EEA) is the environment in which the natural selection of its distinctive biological traits has taken place. For all species, except humans, this environment needs to be very similar to its current environment, because the more different its current environment from the EEA, the fewer possibilities this species has of survival, since, quite obviously, its traits will no longer be adaptive to the environment where it lives (setting aside now the case of domesticated species). But that is not the situation with humans, and later on we will find out why.

Genetic knowledge is, in metabolic terms, very 'cheap'. All organisms, even the most simple ones, do have some form of genetic knowledge to the extent that they have DNA sequences specifically adapted to replicate themselves in a given environment. But it is a rather 'stupid' form of knowledge: the same stimulus will tend to give rise to the same reaction. Even though not all genetic knowledge is equally stupid, for genes can give rise to a certain degree of phenotypic plasticity that produces different responses in different environments, there is a limit to that plasticity. The problem is that in a changing environment, genetic knowledge can be deadly. Once the environmental characteristics that made a particular reaction adaptive have disappeared, that biological trait, whatever it is, is no longer adaptive and it can very well lead to the extinction of the species unless a new mutation that happens to be adaptive takes place. But in big animals, that is, those with a long reproductive cycle, mutations appear (and, especially, accumulate) very slowly. Hence 'Mother Nature' (that is, natural selection) has provided these big animals with another form of knowledge. This is another form of knowledge that cannot change the somatic characteristics of the species but it can change its behaviour and make it adaptable to the new circumstances; namely, it changes the brain instead of changing the (rest of the) body. This is what I call 'individual knowledge'.

Individual knowledge

Individual knowledge is information from the environment that may give rise to a change in the behaviour of the individual animal without changing its DNA, that is, without having to wait for a mutation in its genome that happens to be adaptive to the new circumstances. Taken at face value, this is a rather simplistic dichotomy, for gene expression can also vary with the environment without having to wait for a mutation. This is what is known as epigenetic variations. But again, there is a limit to this variability. Hence, we need some additional mechanism of adaptation that goes beyond that provided by genetic knowledge. On the other hand, it is true that all organisms with an equivalent genotype are likely to follow different developmental trajectories in different environments. But this does not necessarily mean that they 'learn' from that environment. Individual knowledge refers to the kind of associative learning that can be found in the majority of vertebrates. To this effect we need to endow the animal with a complex nervous system that is not only capable of processing information and generating the appropriate reaction, according to its genetic knowledge, but that is also capable of storing that information, in such a way that when new information comes in, it can be processed not only through its genetic knowledge but also through the previous information that it has stored and, consequently, generate a new form of behaviour.

Imagine a land-living organism in a place where there is very little water, a desert, for instance. All of a sudden, there is a climate change, as has been the case quite often in ancestral times, and it starts to rain so that several pools of water are formed: lakes, rivers, etc. Since those pools of water have never been there before, there is nothing in the genetic knowledge of that organism that provides the appropriate information for this new environment: 'if you try to walk on those pools you will sink and will drown, since you cannot breathe under water'. Even if the first time the animal tries to walk on the water and sinks it manages to get out and survive, as no information has been stored anywhere in its nervous system as to the dangerous nature of that place, the next time it comes across a pool of water, instead of avoiding it, the animal will do the same thing, and the same applies to all the members of that species until a mutant is born that, instinctively, avoids the pools of water. But if that mutation does not come out in time it may well be the case that the species becomes extinct because all its members have drowned. But suppose now that that organism has some possibility of storing that information, the information according to which it cannot walk on top of the water and, if it sinks, it is unable to breathe. In that case, the behaviour of the animal can change by not trying to walk on top of the water without having to wait for any change in its genome. That information that the organism has been able to store somewhere in its nervous system that will allow it to learn from the environment, that is, to adapt its behaviour to a changing environment without changing its genetic knowledge, is what I term 'individual knowledge'.

This grossly simplistic example throws into relief the main attribute of individual knowledge, namely, knowledge acquired by an individual organism as a result of its interaction with the environment. Since those interactions will be different in different individuals from the same species, the individual knowledge that each will have will also be different, and the behaviours to be derived from those, different, individual knowledges will also be different, despite the fact that, as members of the same species, they all have the same genetic knowledge. Let us take a quick look at the key characteristics of this new form of knowledge:

1. It is not innate. Unlike genetic knowledge, organisms are not born with individual knowledge but acquire it through their interactions with the environment.
2. It can be accumulated but only in the individual nervous system. To this effect, the individual needs to have a complex nervous system with the capacity to

store and process information coming from the environment. In other words, it needs a brain with long-term memory. The majority of vertebrates have some form of individual knowledge, but the bigger their brains (the more complex their nervous systems), the more individual knowledge they will be able to accumulate. Birds and mammals are the species with the largest brains and, hence, with more possibilities of accumulating individual knowledge.
3. It is not shared, by definition, it depends only on the experiences of the individual, hence different individuals from the same species but with different experiences will have different individual knowledges.
4. And it is not automatic. It needs to be duly processed by a cognitive system, that is, by a complex brain. So it is not as cheap in metabolic terms as genetic knowledge. It is more expensive since the animal needs to have special organs devoted to the storing and processing of the information that produces individual knowledge.

We can see that the main advantage of individual knowledge is its rapid adaptability. Animals can change their behaviour according to circumstances, according to a changing environment without having to wait for a mutation. Thus, it is very good for big animals with long reproductive cycles. They cannot wait for the right mutations since these appear only very sparsely, which can be lethal in situations of rapidly changing environments. Its main disadvantage, however, is that it is (metabolically) expensive. We need special organs to process individual knowledge, we need a big brain with long-term memory, so small animals (invertebrates, etc.) are likely not to have it or to have very little of it. Furthermore, because it is so expensive metabolically, in stable environments animals are likely to turn individual knowledge into the, cheaper, genetic knowledge. If a mutation appears that makes land-living animals automatically avoid pools, without having to have had any bad experience (sinking, drowning etc.), this mutation is likely to spread very rapidly so that what was initially individual knowledge is turned into genetic knowledge (this is what biologists call the 'Baldwin effect').

These two forms of knowledge have been the determinants of animal behaviour for much of the history of life on this planet. This was so until the arrival of a new species, the human species, which is the only species that has been able to add substantially to those two forms of knowledge – genetic and individual – a new one. This is the distinctive and unique form of knowledge of the human species (with some qualifications that I have already pointed out) that I call 'cultural knowledge'. But before analysing the characteristics of this new form of knowledge it is time to say a few words about the origins and evolution of our species, for only by looking carefully at the nature of this process will we be able to understand those characteristics.

In Figures 1.2 and 1.3 we have, in a nutshell, the evolutionary history of the human species. The process that goes from our last common ancestor with the chimpanzee, six million years ago, until the appearance of our species, Homo sapiens, is known as the process of human evolution. Our objective is to find out what happened during this process of human evolution that gave rise to the emergence

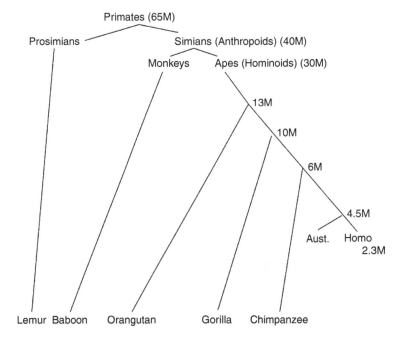

FIGURE 1.2 Primate evolution

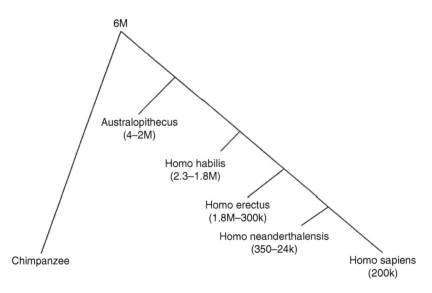

FIGURE 1.3 Human evolution

of this very distinctive characteristic of humans: our capacity to acquire culture, cultural knowledge. And what is 'cultural knowledge'?

Cultural knowledge is the knowledge that an organism (which can only be, with all due qualifications, a human individual) acquires from the interaction not with the environment (individual knowledge) but from other organisms, which are normally other human individuals. Humans can do this to a far greater extent than any other animal. The same as individual knowledge, cultural knowledge is learned by means of our brains. So there must be something very distinctive in the human brain that makes it very different from the brain of all other species. And what could that possibly be?

The growth of the human brain

The human brain is a really complex organ. It has 86 billion cells called neurons (86,000,000,000). Its size is, on average, 1,330 cc and its weight is 1,300–1,400 g, which is two per cent of total body weight (of an adult human being), but it uses up 20 per cent of all the energy consumed by the body. So it is a very expensive organ. Certainly, the first thing that catches our attention when we compare the human brain with the brains of other animals, especially our closest relatives, is its size. We have much bigger brains than the other primates. For instance, our brain is three times as big as the brain of a chimpanzee.

We do not have the biggest brains in the animal kingdom, though. Whales and elephants have bigger brains than us. But they also have bigger bodies. There is a direct relationship between the size of the body of an animal and the size of its brain, so that the bigger the body the bigger the brain. This relationship, however, is not regular because bodies grow 'faster' than brains. This means that bigger animals have, proportionately, smaller brains than smaller animals. Thus, the brain size of a shrew is ten per cent of its body, so it is five times 'bigger', in relation to its body size, than the human brain. This sort of relationship, between the size of the brain and the size of the body, is called an 'allometric' relationship, that is, a relationship that can be expressed by means of a mathematical formula: if we know the body size we can know the brain size and vice versa.

Once we know that, we are in a position to understand what is the encephalisation quotient (EQ), which is the amount of brain size that surpasses the size it should have according to the allometric relationship we have just seen. In other words, it is the amount of surplus cerebral mass that we have in relationship to the size that corresponds to our body weight. Roughly speaking, the EQ is the measure of the intelligence of the animal. It should be noted that there is no relationship (that has been recoded so far) between EQ and cognitive capabilities between members of the same species. The relationship is only obtained between members of different species. Now, in what concerns encephalisation, humans certainly have the biggest EQ in the animal kingdom, that is, we have a lot of surplus cerebral mass in relation to the size of our body. (Dolphins, however, have an EQ very similar to humans.) In this sense, we can safely conclude that human brains are certainly very large. The question is why; why are they so big, taking into account that they are metabolically so expensive?

Figure 1.4 shows the growth of the brain size during the process of human evolution. We can see very clearly a sharp discontinuity that began to show itself

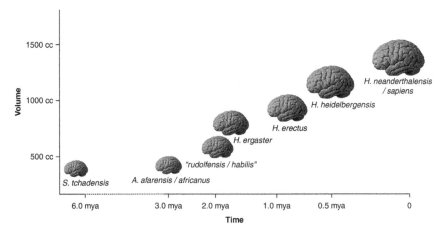

FIGURE 1.4 Growth of the human brain

approximately 2.1–2.3 million years ago. Until that time, the brain of our ancestors did not grow much bigger than the brain of a chimpanzee, which is approximately 400 cc. But after that, which was precisely when the genus homo appeared, it started to grow at a relatively fast rate (in evolutionary terms). So, during the six million years of human evolution, our ancestors had for almost four million years a brain the size of a chimpanzee, but then in a bit more than two million years it grew three times as much. To some extent, this was due to the parallel growth of their bodies: Australopithecines were also smaller than modern humans. However, our brain has increased in size much more than our body size.

Riddles of human evolution

It seems that roughly at that time, between two and 2.5 million years ago, the average temperature on earth started to cool down. As a result of this process the rain forests that existed in the south-east of what is now the Sahara desert disappeared and became the savannahs that we find there now. All the animals that were adapted to live in the rain forest retreated towards warmer and wetter regions, to forest refugia, and among these animals were the primates, the ancestors of the chimpanzees and gorillas that we find nowadays in the rain forest of central Africa (south-west of the Sahara desert). Other primates remained in the savannah, such as baboons and many other monkey species adapted to more open habitats. This is normally the way new species are created: there is a climate change and a particular group of animals is no longer adapted to the new environment, the majority of these animals emigrate except for a small group that for whatever reason gets isolated from the rest. New mutations appear that might be adaptive in the new environmental conditions, so these mutations will not replicate among those who have emigrated but will in the isolated group. After a few generations we will have a new species.

But sometimes the right mutations are hard to come by, especially in large mammals. That was the case for a small group of Australopithecines that, for whatever reason, also found themselves isolated in the savannah. Life for that isolated group of Australopithecines must have been very tough. Even though they were already bipeds, they spent a good amount of time in trees, as the majority of primates do, from which they obtained food (vegetables and fruits), shelter and protection from predators. The savannah was for them, with very few trees around, a rather hostile environment. Bipedalism probably turned out to be more adaptive in the savannah (because their bodies were less exposed to the sun), and that would explain why the Australopithecus survived there, whereas other primates which perhaps did not manage to get into the rainforest probably became extinct. But they still had the problem of food. Australopithecines were vegetarians, or mostly so, but very few of the vegetables they were using as food could be found in the new environment.

Both gorillas and chimpanzees are vegetarians, but whereas gorillas are 100 per cent vegetarians, chimpanzees eat meat from time to time – they actually hunt monkeys and small animals. So it is not totally unconceivable that some of those isolated primates started to eat meat, first as scavengers and eventually from their own hunting. The problem is, however, that the body of great apes has not been naturally selected either for eating meat or, *a fortiori*, for hunting. Great apes in general do not have powerful claws or teeth with which to kill their prey and tear off their flesh, they are not fast runners (practically all animals run faster than them) and, on top of that, their digestive apparatus has not been naturally designed to take in meat, at least, not systematically – meat is pretty easy to digest, though, only a small gut is needed whereas tough vegetable matter needs a more specialised gut. In any case, non-adaption for meat consumption is what explains why chimpanzees only eat meat occasionally. But for our ancestors it was not enough to eat meat only from time to time, they had to eat it quite often because there was very little (vegetarian) food available. How did they manage to solve this problem?

There is evidence that the first meat eating hominid was Homo habilis (2.3–1.8 Mya), who was also capable of producing very rudimentary stone tools – though the oldest stone tool that has been found is three million years old, so most likely some Australopithecines were already able to make some tools. The purpose of these first stone tools was in all probability butchery, that is, to crush dead animals' bones and tear off their flesh. Note that there is a very important difference between using an object as a tool and *producing* a tool. Chimpanzees are quite good at using objects as tools, that is, they use stones to crash nuts, but they are not very good at producing any tool, even the ones which have been specifically trained by humans for this purpose. The tools made by Homo habilis were really very simple. Basically what they did was knapping: smashing one stone against another in order to polish the surface of one of them or to produce stone flakes – small pieces of stone with very sharp edges that could be used for cutting.

No matter how simple this technique might look to us, there is no non-human animal capable of doing this. It is unclear why this should be so. In evolutionary terms, the reason is undoubtedly that they do not have any need to do it; in other words, there were no selective pressures on their ancestors that might have generated this cognitive ability, precisely the selective pressures that turned this capacity into a matter of life or death for early hominids.

The systematic consumption of meat had a very important metabolic consequence for those early hominids: it provided them with a valuable source of proteins and fats with which they could feed their growing brains. As we have seen, a big brain is a very expensive organ. It would have been very difficult to feed this organ only with the proteins and fats that can be obtained from vegetables. So a positive feedback loop started to develop between growth of the brain and the consumption of meat. Note, however, that it is not meat eating by itself that produces bigger brains, for carnivores are not necessarily brainier than herbivores. Meat simply provided the additional proteins and fats needed for brain growth. But the growth of the brain must be explained, like any other biological trait, in functional not causal terms. With bigger brains our hominid ancestors were capable of producing better tools, which enabled them to consume more meat, which in its turn helped them to feed their brains, etc. Brain growth is what enabled our hominid ancestors to generate the appropriate adaptive behaviours in the savannah, such as the production of tools for hunting or butchering, while having a clearly maladapted body to that effect. We shall come back to this.

There is no evidence that Homo habilis did any hunting, so in all probability they were only scavengers. They might have done occasional hunting, as chimpanzees do, but palaeoanthropologists have not been able to prove it. There is plenty of evidence, by contrast, that Homo erectus, which is the next hominid species (1.8M–300 mya), was already a skilful hunter. Homo erectus had a much bigger brain, up to 1,000 cc while Homo habilis reached only 660. They were the first hominid to leave Africa: Homo erectus remains have been found as far away as China and Java. So they managed to colonise all of Europe and Asia. They slept on the ground while Homo habilis still slept in trees – though the evidence for that is open to doubt. They knew how to make fire. Thus, apart from being good hunters, they were also able to cook the meat and hence to digest it more easily. Some argue that they might have had a rudimentary language, but this is just speculation. And on top of that, Homo erectus were able to systematically reuse the tools they made.

This is a very important issue that deserves careful attention. Even though, again, we have no conclusive proof, it seems that Homo habilis would make new tools each time they needed them – which is what chimps do with the stones or little branches they use as tools most of the time (though occasionally they also reuse them). But that was not the case with Homo erectus: they would make tools and keep them in order to use them again when the need arose. The reuse of old tools, together with the systematic production of new ones, is made possible thanks to a cognitive ability that only humans have fully developed: mental time travelling. Humans are constantly mentally travelling through time: we are always worried about what is going to happen to us in the future, even in the distant future, and we keep remembering what we did in the past, or what was done to us in the past, quite often a very distant past. There is a clear relationship between this mental time travelling ability and the production and reuse of tools. I would neither bother wasting my time producing something that will only be useful to me in the future, nor would I keep that very same object once it has been used just in case I might need it again, if I were not able to imagine myself as existing both in the past and at that future time.

But even more importantly than mental time travelling is that this reuse of tools enabled Homo erectus to share the tools they made, so they could not only reuse their own tools but also reuse tools made by someone else. Sharing tools is a crucial milestone in human evolution. In order to use a tool made by someone else you need to have another very important cognitive capacity that only humans have (or, again, only humans have fully developed), and that is the capacity to interpret manufactured objects in terms of the producer's intentions, in other words, the capacity to figure out and to understand intentionality. True, very simple tools can be reused just by trying them out and seeing in them a use for ourselves irrespective of the maker's objectives. But that cannot be the case for more sophisticated artefacts. We shall now look at the origins of this capacity and why it was so important in the history of human evolution.

Theory of mind and the modular structure of the human mind-brain

Consider the way we use manufactured objects in our day-to-day life. How do I know the purpose of any manufactured object, specially an object that I have never seen before? I need to figure out the intentions of its maker. In our day-to-day lives, we are surrounded by manufactured objects. We make use of them all the time, yet we take no notice of what our mind does in order to figure out how to use them. The process consists of placing ourselves into the manufacturer's mind. The fact that non-human animals lack this cognitive capacity, or have it much less developed, is precisely what prevents them from making *and* sharing tools. This capacity to figure out intentions behind events or objects is, in effect, the ability to think that certain things happen or certain objects exist, because someone wants it to be so. This ability originates in a cognitive mechanism, unique to the human species in its most developed form, called theory of mind (ToM).

ToM is the capacity humans have to figure out the existence of mental states (that is, intentions, belief, desires) behind certain objects, behaviours or events (such as movements of an animal or of another human being, or, as we have just seen, manufactured objects) in such a way that the existence of those objects, behaviours or events is explained as a result of those mental states. ToM is also referred to as 'mentalisation' or 'social cognition'. Later we will see the role of ToM – a distinctive cognitive ability that is unique to humans – in the generation of cultural knowledge, which also happens to be unique to humans. True, embryonic versions of our ToM mechanism, together with rudimentary forms of what we call here 'cultural knowledge', can also be found in other non-human species, particularly in primates. But in no other species have they reached the levels of development that can be found among humans.

Be this as it may, let's pause for a moment and return to the evolution of Homo erectus. Why did the Homo erectus brain grow so much (more than twice that of the Australopithecus) in just one million years? It was the result of natural selection acting on the Homo erectus population living in a hostile environment to which its body was not adapted. The fate of any other animal in such an environment would have been either mutation or extinction. But in the case of Homo erectus, and its ancestors, Australopithecus and Homo habilis, natural selection took a different turn: as well as changing its body, it changed its brain. The reason for that is that evolution always makes do with whatever is available, and always with the cheapest solution in biological terms; natural selection eliminates costlier solutions. Undoubtedly, the Australopithecines' bodies also evolved: becoming exclusive bipeds, for example, becoming taller and possibly developing the physiological mechanisms for endurance running. However, in all probability too many mutations would have been needed to make the body of our primate ancestors fully adapted to life in the savannah. But fewer were needed to make their relatively big brain able to handle this task. Note that primates already have a very big brain as compared with other mammals; increased brain size can be considered a distinctively primate way of adapting. And how can the evolution of the brain make up for the (relative) non-evolution of the body?

Let's compare the brain of any animal with a computer. We normally think that what makes a computer powerful is the capacity of its hard disk. But a hard disk without any software is totally useless. What makes a computer powerful, that is, capable of processing texts, images, videos, graphics, etc. are the specific programmes uploaded onto it. The hard disk capacity, namely its memory, makes sense not because it enables the computer to accumulate lots of information (that is, data). Rather it is powerful when it can accumulate lots of programmes; programmes are what makes the computer powerful. The same thing happens with the brain of any other animal. Big brains are those that store lots of 'programmes' rather than information. These programmes that we have stored into our brains, and that enable us to manipulate the information we get from the environment, are called cognitive modules.

In very plain language, a cognitive module is a programme that enables us to process a certain kind of information and to generate an appropriate reaction or behaviour: much the same as a computer programme. A computer without a text processor, for instance, cannot produce a text, just as a computer without an image processor cannot produce an image. You may upload images onto the computer, but nothing will appear out on the screen, simply because it does not have the suitable programme to process that information. A similar argument could be applied to our brain: we are able to process particular kinds of information coming from the environment because we have the appropriate programmes, that is, cognitive modules. We should emphasise, however, a very important way in which a cognitive module is different from a computer programme. Cognitive modules have much greater functional plasticity than software. The actual domain of a cognitive module, that is, the set of behaviours that it can generate at any particular place and time, is normally bigger than the behaviours for which it was naturally selected (its 'proper domain'). That could not be otherwise if we take into account that a cognitive module has not been produced by an engineer but rather by natural selection. Note, on the other hand, that the cognitive modules' behavioural flexibility is not necessarily a shortcoming, for it can very well provide adaptive advantages in changing environments.

This is why the human brain grew so much. As I have already pointed out, primates have bigger brains than other mammals because selective pressures tend to act upon their brains rather than on (the rest of) their bodies. Human evolution followed the same path, but in an exaggerated form. The human brain had to accommodate many more cognitive modules than the brain of any other animal. And that was so because without them our ancestors would not have been able to survive in the environment where they found themselves more than two million years ago.

So far I have said nothing about cultural knowledge. Is the growth of the brain not in any way related with the emergence of cultural knowledge? Traditionally it was thought that humans had such big brains as compared to other animals because we had to accommodate cultural knowledge, a form of knowledge that only humans possess. But this is unlikely for a very simple reason: the human brain started to grow well *before* cultural knowledge was in any way significant, that is, well before humans had language (see Figure 1.4). This also means that once cultural knowledge became indispensable for human survival and reproduction, many of those cognitive modules might have played a crucial role in the very process of cultural learning, such as our ToM or our capacity to learn a language (see below). Consequently, their existence in modern humans has to be accounted for by that new function as well. The point I wish to make, however, is simply that the initial growth of the human brain cannot be the result of the need for cultural learning, for it is pointless to have such a big, expensive organ when there is no, or very little, culture to be learned.

The modules grew in our brain through the process of natural selection: random mutations in the genes that code for our neural architecture (the structure of our brain) that turned out to be adaptive, that is, they generated behaviours that enhanced the individual's reproductive fitness in a particular environment. Natural selection is the same process that changes the body of any species of living organism and turns it into a new species. The difference in our case is that instead of changing our bodies (or apart from changing our bodies a little bit as well, since the body of a Homo sapiens is certainly different from that of Homo erectus or Australopithecus) it changed our brains, making them more and more powerful, equipped with more and better cognitive modules, so that we could compensate for our maladapted, or not fully adapted, bodies with adaptive behaviour. Thus we know the answer to the question as to why the human brain grew so much: because adaptation for our hominid ancestors was what we might call behavioural adaptation instead of (exclusively) bodily adaptation. By behavioural adaptation I mean that adaptation does not originate in particular characteristics of the body, such as claws, bodily hair, teeth length, etc., but in the performance of appropriate behaviours. And big brains made it possible to generate those appropriate behaviours.

Four very important principles should be taken into account when we talk about cognitive modules:

1. They are part of our genetic knowledge. We do not get them from the environment or from interaction with the environment. Rather, we get them from our ancestors. We are born with them, or they grow inside us as we mature ('maturational knowledge', in the terms of American philosopher Robert McCauley). In other words, a newborn baby does not have all the cognitive modules that we find in a fully developed adult brain. But he or she will 'grow' those modules with a certain stimulus coming from the environment. This means that what the environment provides is not the module itself but rather just the stimulus to make it grow. It is as if the baby was born with the seeds and we only need to water those seeds and feed them to make them grow and turn them into plants.
2. From this point of view, a strictly cognitive one, the difference between the human brain and that of non-human animals, especially our closest relatives (chimpanzees, gorillas, monkeys, etc.), is purely quantitative: we have 'more' and they have 'less'. Our brain is larger because our computer has more programmes than theirs. Later on we will see that this is not the only difference; there is also a qualitative difference.
3. Cognitive modules are what cognitive scientists call 'domain-specific'. Each module is only capable of processing particular kinds of information, once again, just like computer programmes. But, as has already been pointed out, here the computer analogy can be somewhat misleading. Whereas a computer programme has been designed by a computer engineer, cognitive modules result from natural selection. And natural selection is a much less refined manufacturer than human engineers. This does not mean that natural selection cannot produce extraordinarily well adapted structures. It means that our cognitive modules might be more flexible in terms of the sort of information they can process than computer programmes.
4. Cognitive modules are functions of the brain, not regions of the brain (even though some of them correspond to particular regions in the brain).

What are some examples of cognitive modules? How many cognitive modules do we have? If we search for 'cognitive module' in Wikipedia we get the following results: the modules that control your hands when you ride a bike, allowing you to keep your balance by turning slightly from side to side; the modules that allow a basketball player to accurately shoot the ball into the basket by tracking ballistic orbits; the modules that recognise hunger and tell you that you need food; the modules that cause you to appreciate a beautiful flower, painting or person;

the modules that make humans very efficient in recognising faces, already seen in two-month-old babies; the modules that cause some humans to be jealous of their partners' friends; the modules that compute the speeds of oncoming vehicles and tell you if you have time to cross; the modules that tell you to look right and left before crossing the street; the modules that cause parents to love and care for their children; the libido modules; the modules that specifically discern the movements of animals; and the fight or flight reflex choice modules.

Some of them are probably distinctive of the human species; others will be shared (totally or partially) with other species. In fact, we do not know yet exactly how many modules we have, and what they are. In any case, here we are specifically interested in the human modules, those that only humans have. I shall say a few words about two of the most distinctive human cognitive modules: the language module and the ToM module. The language module is a clear example of a distinctively human competence. It is often called the Language Acquisition Device, which is the name given to it by American linguist Noam Chomsky, who was the first to postulate the existence of this specifically human cognitive ability. The language module appears to correspond to a particular region of the brain, the so-called Broca and Wernicke areas (in the left hemisphere). People who have injuries in that particular area may have impaired linguistic abilities, while the rest of their cognitive capabilities will remain intact. How did Chomsky reach the conclusion that we have a language module? I have just said that cognitive modules are part of our genetic knowledge. But language appears to be learned rather than innate. Newborn babies cannot speak at all; they have to learn how. And depending on which country they are born in they will learn one language or another. So how could there be a language module, considering that language is clearly something we learn rather than something with which we are born?

The answer lies in the fact that the language module is not a language but the *capacity* to learn a language. Chomsky realised that all human languages are extremely complex from the logical point of view. Think for a moment about syntactic rules and how complex they are in all languages. However, little children in all societies are capable of learning a language in a relatively short time and with very little instruction. This is what Chomsky called the 'poverty of the stimulus' hypothesis. Traditionally it was thought that children would learn a language simply by listening to adults' speech, repeating the sentences they heard and inferring from them (by means of an inductive and unconscious process) the grammatical rules that preside over the production of those sentences. But this is unlikely, Chomsky concluded, because the information children obtain from adults is very poor and the grammatical rules of any language are extremely complex. So in order for children to be able to learn a language, they must have a special device in their brains that predisposes them to learn a language. This device must *already have* the most basic and elementary grammatical rules of that language. This is something that can only be found in human brains, since no other animal is capable of learning a human language, not even when they have been explicitly taught. Chomsky termed this mechanism the 'Language Acquisition Device' (LAD). We do not carry a language per se in our genetic knowledge, but rather the capacity to acquire a language. And this capacity includes the most basic grammar rules, because all human languages share them, what is known as Universal Grammar or Deep Grammar. We do not actually 'learn' a language: we 'grow' a language inside our minds with the help of an external stimulus: the spoken words that children hear while they are growing up. (That explains why people who do not hear spoken language as children are incapable of speaking a language later, or they find it very difficult.) Later on we will see the problems that arise when we try to account for the evolutionary origins of this LAD.

The LAD is the clearest example of a cognitive module. Linguistic knowledge is clearly maturational knowledge. We obtain it as we mature in the appropriate environment, but we are already born with the seeds of that knowledge. We have already alluded to another cognitive module: the ToM. Our ToM is our capacity to figure out the mental states behind people's actions, behind the production of manufactured objects and behind any event that appears to be the consequence of somebody's actions. This is an extremely important cognitive module. Like the LAD, it is distinctive of the human species. But it is more basic than the LAD, since we cannot have language without having a fully developed ToM. What does 'understanding someone's speech' mean if not figuring out his or her intentions when speaking? I can understand a language, which is just a series of sounds, when I am able to imagine the intentions – the mental states – of the person who pronounces them.

Autism is a mental disorder that is normally associated with a defective ToM. Even though autism manifests itself in different degrees, those who suffer from the most severe forms of it are incapable of learning a language, and the reason appears to be because they have problems with their ToM module. We have not been able to identify the exact location in the brain of the ToM, even though there are some hypotheses. The discovery of the so-called mirror neurons in macaques by the Italian neuroscientist Giacomo Rizzolatti suggested a promising venue for that research. These are neurons that fire both when the animal performs an action and when the animal sees the same action performed by another; which somehow suggests that the animal has the capacity to imagine the other's intentions. But those neurons were initially discovered in macaque monkeys, which do not seem to have a particularly developed ToM. Autistic children, on the other hand, with a clear deficit on that cognitive function, do seem to have their mirror-neuron system impaired, together with other neuron systems. So it is unclear which of those alterations is responsible for their disorder. And, as I have said, cognitive modules are functions of the brain, not regions, so they do not have to correspond to specific areas of the brain. Rather our ToM may be produced by a complex set of neural interconnections.

Developmental stages of Theory of Mind

Whatever the exact brain location of ToM, developmental psychologists have discovered that, like our ability to speak, ToM grows inside our brain in different stages. The first and most basic one, which manifests itself only a few months after birth, is gaze following. Only children suffering from severe autism are incapable of following another person's gaze; this is in fact one of the most unambiguous symptoms of autism. It seems that chimpanzees can follow a gaze, and probably dogs too. Notice that to follow another person's gaze, you need to understand that the person is seeing something that might be of interest to you, in other words, that that person has perceptions the same as you – we do not follow the gaze of a statue. This is not exactly a ToM but it comes near to it – you are imagining that the other has perceptions, but not yet fully-fledged mental states. Another gesture symptomatic of the existence of a ToM and which, again, it seems that only humans can perform and understand is pointing. By nine to 18 months of age, all healthy infants can understand what pointing is. Pointing is a little bit more complex than just gaze following. To understand someone pointing at something you need to understand that the person wants to tell you something, that is, that he or she has a mind with intentions and that those intentions might be of interest to you. Pointing represents an early stage of ToM. The full development of ToM only comes when the child reaches four to six years of age. At this point, the child already has remarkable linguistic skills and becomes capable of passing the so-called 'false belief task'.

In this task, a researcher shows two boxes to a child and places a sweet in one of them. A confederate, who has also been present, leaves the room. While he or she is absent, the researcher moves the sweet from one box to the other and then asks the child where the other person will think the sweet is. Children under four will give the wrong answer: they will say that the other person thinks that the sweet is where it actually is. They are not able to understand that someone might entertain 'false beliefs'. By contrast, children over four will give you the right answer. They will know that the confederate does not know where the sweet actually is and that he or she has a false belief. This is important because it shows that at this stage children can represent what we call 'epistemic mental states'. These are representations of states of affairs that can be different from what they represent. In other words, they can understand what a 'mind' is. They can understand that people have 'mental states', that in those mental states there are beliefs about the world, which are representations of some bit of reality, and that those beliefs can be different from the reality they are meant to represent, that is, they can be false beliefs.

Researchers argue as to the precise timing of ToM development: is it really at around four years of age, or can younger children have some rudimentary mentalising capacities? It is difficult to determine whether this cognitive ability exists in very young children and in non-human primates because most tests entail linguistic communication. The few tests that do not require communication through language imply that some mentalising capacities are present in much younger children (as young as 15 months) and also in some non-human primates. Be that as it may, only humans have a fully developed ToM, even if some other species might have embryonic versions of it.

Evolutionary origins

What about the evolutionary origins of our ToM? What could be the selective advantage of having a ToM in the particular environment where human evolution took place? Obviously, we will never know for certain, because we are talking about something that took place millions of years ago. But there are some theories that try to explain how ToM could have enhanced our ancestors' reproductive fitness.

The first is that ToM facilitated cooperation. Naturally enough, if you have to work in close cooperation with someone else, it is quite useful to know his or her intentions, thoughts and beliefs. Even though we do not know for sure what cooperative activities our ancestors engaged in, probably one of the most important ones, particularly from Homo erectus onwards, was hunting. Humans, and hominins, are not and were not 'natural' hunters. In fact, the human body is very poorly endowed with hunting devices (fangs, claws, etc.) compared to other carnivores. And for hunting big animals they needed not only tools (that is, weapons, such as stone axes and spears) but also help from other hunters. In present-day hunting-gathering societies, which are the societies that probably most closely resemble our hunting-gathering ancestors, hunting is almost always a collective activity. And the same applies to chimpanzees; their occasional hunts are always carried out in groups. Another reason for making the hunt a collective activity is that the meat of a big animal cannot be consumed by one individual (and before the discovery of salt it could not be preserved either), so it has to be shared. And if you have to share what you hunt with other people, they might as well give you a hand. Some scholars have criticised this theory on the basis of sex. In nearly all documented hunting-gathering societies, and among chimpanzees, hunting is a male activity. This would imply that ToM developed first in males, since males engaged in the cooperative activity for which ToM came to be very useful. But males the world over seem to be less gifted at grasping other individuals' mental states than females. So if ToM developed initially out of a predominantly male activity, it is hard to understand why it is that females have better ToM abilities than males.

The so-called 'attachment theory', according to which ToM developed first out of the close interactions between a mother and her child, solves that problem. Due to the fact that humans, or pre-humans, had big brains, they had to be born with their brains still underdeveloped; otherwise, they would not have been able to pass through their mothers' pelvic canal. Human offspring are born very early in biological terms, especially in terms of brain development, and, consequently, they are far more vulnerable than any other animal's offspring. As a result, they need to be looked after for much longer than other mammals, and mothers are mostly likely to be in charge of this prolonged care. A child shares 50 per cent of his or her genetic endowment with the mother, the father and any full siblings (children of the same parents). But unlike the father or full siblings, the mother is the only individual who can be 100 per cent sure of who her children are. For prolonged care to be successful, a close interaction between the mother and her child is needed, and ToM seems to be well suited to this task. True, all female mammals care for their offspring and not all of them have developed a ToM. But only human neonates need a great deal of extra care due to their characteristic vulnerability. That would explain why humans developed a ToM, especially human females. It would also explain why women the world over are much better than men at grasping other people's intentions and mental states in general, and the fact that autism (ToM malfunction) is overwhelmingly a male disorder.

A third factor that might explain the evolution of a ToM in humans is the need to share tools. We have already seen that humans are the only animals that can both make tools and share their use. In order to share a tool that has been made for a particular purpose, especially if it is a rather sophisticated tool, we need to be able to figure out the purpose of that tool; in other words, we need to be able to figure out the intention behind the production of that tool. And for this we obviously need a ToM.

The ToM module is extremely important for human cognition. It is also extremely relevant for explaining the uniquely human capacity to produce cultural knowledge. But before seeing why ToM is so important for our purposes, let us review what we have seen so far in our short journey through human evolution. The growth of the human brain is the most distinctive aspect of human evolution and the reason it came about is clear: increasing brain size is the typically primate way of adapting. Humans merely made use of that resource much more effectively and extensively than any other primate, which enabled them to accommodate an impressive number of programmes, that is, cognitive modules, in their oversized brains. From this point of view, as we have already seen, the difference between humans and non-human animals, or between human brains and the brains of non-human animals, is merely a quantitative difference: we have more and they have less. According to this interpretation of human evolution, it is clear that the human brain did not grow to 'accommodate culture', as it was believed some time ago, but to accommodate 'instincts', that is, genetic knowledge. That is why, as American psychologist and philosopher William James pointed out long ago, humans have more instincts than non-human animals, not fewer.

The growth in our ancestors' brains made room for this particular form of genetic knowledge that constitutes our cognitive modules. We still do not know for certain how many of these cognitive modules we have. But in all probability ToM is among them. Whatever the initial function of that module (to boost cooperation, facilitate interaction between a mother and her offspring or aid the making and sharing of tools), it had a remarkable side effect. It was an absolutely exceptional, unique by-product that turned the so far quantitative difference between humans and non-humans into a qualitative difference. And that was the appearance of cultural knowledge. How could culture originate out of our ToM? We will address this issue in the next chapter.

Bibliographical note on Chapter 1

There are many introductory biology texts with excellent summaries of the basic principles of evolution. Dawkins's (1976) classic text, perhaps one of the best known popular science books, contains a very clear description of the gene-centred view of evolution, which happens to be the dominant approach in contemporary evolutionary biology. In Dawkins (1986) we have an excellent development of the ideas that were put forward in 1976. Matt Ridley (1999) could also be a useful complement to Dawkins. For those interested in the wider theoretical and philosophical implications of evolutionary theory, a look at Dennett (1995) is essential. A different view of evolution, rather critical of the gene-centred perspective, can be seen in Jablonka and Lamb (2006). This is not an easy read for beginners but it is comprehensive and well argued. Regarding human evolution, Coolidge and Wynn (2009) provide a good interdisciplinary review, which includes perspectives from archaeology, paleoanthropology and neuroscience. Dunbar (2005), Gazzaniga (2008) and Suddendorf (2013) are less comprehensive works specifically centred on an account of human uniqueness. In terms of the human mind and its evolutionary origins, a good classic text is Donald (1991), though it presents a rather specialised theoretical approach. More general perspectives on the modular theory of the mind can be found in Carruthers and Chamberlain (2000) and Hirschfeld and Gelman (1994). But the landmark text on this issue is still the work of American philosopher Jerry Fodor (1983). Another very important text on the modularity of the human mind, specifically focused on the cognitive foundations of culture, is Barkow et al. (1992). These authors are very critical of perspectives from the social sciences. Mithen (1996) looks at the evolution of the mind from an archaeologist's point of view, with some critical observations as regards mainstream modularity theories. A good cognitive analysis of the human capacity for language can be found in Pinker (1994); it is comprehensive and well written, though beginners might find some chapters hard to follow. A work that is more specialised but full of deep theoretical insights about human cognitive uniqueness is Deacon (1998), which contains an interesting approach to the problems of the origins of language different from the standard Chomskian perspective espoused by Pinker. The study of ToM or mentalisation has given rise to a plethora of specialised literature; Carruthers and Smith (1996) and Baron-Cohen et al. (2000) provide good introductions. The relationship between parenting and human uniqueness, specifically related to humans' mentalising faculty, is very well analysed in Hrdy (2011).

References

Barkow, J.H., J. Tooby and L. Cosmides. 1992. *The Adapted Mind: Evolutionary Psychology and the Generation of Culture*. New York: Oxford University Press.
Baron-Cohen, S., H. Tager-Flusberg and D.J. Cohen. 2000. *Understanding other Minds: Perspectives from Developmental Cognitive Neuroscience*. Oxford: Oxford University Press.
Carruthers, P. and A. Chamberlain, eds. 2000. *Evolution and the Human Mind*. Cambridge: Cambridge University Press.
——— and P.K. Smith, eds. 1996. *Theories of Theories of Mind*. Cambridge: Cambridge University Press.
Coolidge, F.L. and T. Wynn. 2009. *The Rise of Homo Sapiens*. Malden: Wiley-Blackwell.
Dawkins, R. 1976. *The Selfish Gene*. Oxford: Oxford University Press.
———. 1986. *The Bind Watchmaker*. London: W.W. Norton.

Deacon, T. 1998. *The Symbolic Species: The Co-Evolution of Language and Brain*. New York: W.W. Norton.
Dennett, D. 1995. *Darwin's Dangerous Idea. Evolution and the Meanings of Life*. New York: Simon and Schuster.
Donald, M. 1991. *Origins of the Modern Mind*. Cambridge, MA: Harvard University Press.
Dunbar, R. 2005. *The Human Story. A New History of Mankind's Evolution*. London: Faber and Faber.
Fodor, J. 1983. *The Modularity of Mind*. Cambridge, MA: MIT Press.
Gazzaniga, M.S. 2008. *Human. The Science Behind What Makes Us Unique*. New York: HarperCollins.
Hirschfeld, L.A. and S. Gelman, eds. 1994. *Mapping the Mind. Domain Specificity in Cognition and Culture*. Cambridge: Cambridge University Press.
Hrdy, S. 2011. *Mothers and Others. The Evolutionary Origins of Mutual Understanding*. Cambridge, MA: Harvard University Press.
Jablonka, E. and M.J. Lamb. 2006. *Evolution in Four Dimensions*. Cambridge, MA: MIT Press.
Mithen, S. 1996. *The Prehistory of the Mind*. London: Thames and Hudson.
Pinker, S. 1994. *The Language Instinct*. New York: HarperCollins.
Ridley, M. 1999. *Genome. The Autobiography of a Species in 23 Chapters*. New York: HarperCollins.
Suddendorf, T. 2013. *The Gap. The Science of What Separates Us from Other Animals*. New York: Basic Books.

2
A NEW FORM OF KNOWLEDGE

Culture in human evolution

Recall our definition of cultural knowledge: the knowledge that an organism acquires from interaction not with the environment but with other organisms. Humans are the only organisms that produce this kind of knowledge on a massive scale – even though, as is normally the case with allegedly unique human characteristics, incipient forms of cultural knowledge can also be found in other species. Now the difference between cultural knowledge and individual knowledge is not always as clear-cut as it might seem at first sight.

Think of the culturally constructed environment of towns and cities. All sorts of implicit messages can be produced by these environments that are likely to influence the behaviour of their inhabitants in many different ways. For instance, our understanding of space and time is in all probability determined to a substantial extent by the place where we live, the distances we have to travel on a regular basis and the means of transport at our disposal. If all these are constraints on behaviour resulting from the interaction between individual and his or her environment, there does not seem to be much of a difference, as far as that interaction is concerned, between 'cultural' and 'natural' environments. A parallel situation has been observed by evolutionary biologists among non-human organisms in what is known as 'niche construction'. Living organisms can alter their environments in such a way that those alterations have an influence on the behaviour of subsequent generations. In a way, there seems to be a form of cultural knowledge behind the production of that behaviour in so far as it originates not in the interaction of the individual with its natural environment but, indirectly, with the other individuals that have altered that natural environment in significant ways. But notice how different that interaction could be from, say, that between a human being and a manufactured object. In niche construction in general, as in the particular case of culturally constructed environments, individuals do not need to grasp in any way the intentions embodied in the environment in order for that environment to have an influence on their behaviour.

That is precisely the key difference between individual and cultural knowledge. Cultural knowledge is always intentional knowledge, namely, information contained in an individual's mind (or in the products of that mind, such as manufactured objects). Philosophers use the word 'intention' to refer to mental representations in general, and this is the way I am using that word now. To assimilate cultural knowledge as such, we need to grasp the intentions embodied in it. ToM's role in the emergence of cultural knowledge is therefore easy to understand. Our ToM enables us to enter, so to speak, other people's minds to figure out their beliefs, desires and thoughts (i.e. their intentions in the philosophical sense). In other words, thanks to our ToM we can appropriate another subject's individual knowledge – the contents of his or her mind. But because that other subject also has a ToM, the sort of knowledge that we find in his or her mind is likely to include the knowledge of other individuals, who in turn will also include knowledge of still more individuals. Thus the knowledge we can acquire thanks to our ToM, which is knowledge that originates in interaction with other individuals, can very rapidly become cumulative knowledge (similar to genetic knowledge in that respect). This is because by acquiring another individual's knowledge we actually acquire all the knowledge that this individual has in turn acquired from other individuals, who also acquired knowledge form others.

An important distinction should be made at this point, and that is the distinction between cultural knowledge and social knowledge, or cultural learning and social learning. Similar to cultural knowledge, social knowledge is also the knowledge acquired by an animal from its conspecifics. Different non-human animals can learn from their conspecifics, especially mammals. It could be argued that this kind of knowledge is different from what I have defined here as individual knowledge in so far as the knowledge is obtained from other subjects and not from the environment. However, the differences are even more remarkable between social knowledge and cultural knowledge. In social learning animals copy each other's behaviour without entering into each other's minds or, at least, not necessarily. An animal might find useful what another one does, such as cracking a nut with a stone, without grasping the intentions behind that act, in other words, without using any ToM mechanism. The difference is certainly subtle at the phenomenological level: how do I know whether an animal that copies an action from another one is learning from the behaviour or learning from the mind? But despite superficial similarities the crucial difference lies in the long term effects of both types of learning. Only the learning from intentions, that is, cultural learning, permits the accumulation of knowledge. Correspondingly, the differences between individual knowledge and social (non-cultural) knowledge are also superficial. By learning from the behaviour of another animal that behaviour is actually treated as part of the environment.

A simple example will show very clearly this cumulative capacity of cultural knowledge and hence the different effects of social learning and cultural learning. While you are reading this text, you might think that thanks to your ToM you are able to absorb into your mind whatever you think I have in mine; our minds are, as it were, in communication and whatever is in mine goes into yours. But the knowledge I am transmitting to you is not only my knowledge. I obtained this knowledge from other people (through books, lectures, etc.), who in turn, got it from other people. So the knowledge you are acquiring from me at this very minute is not in any conceivable way 'individual' knowledge; it is inherently collective. In other words, by just reading this text, or reading any other text, you are acquiring knowledge produced by thousands and thousands of individuals. This is not a miracle; this is what cultural knowledge is all about.

That is the reason why many authors think that cultural knowledge is like a virus that colonised human (or pre-human) minds and found there the appropriate conditions for its reproduction – an interesting metaphor, as we shall see. But why do we call it a 'virus'? A virus is a nucleic acid sequence that gets into our body and reproduces by making our body produce copies of the virus. But not because our bodies were designed by natural selection to lodge viruses; rather viruses are merely a side effect (an unfortunate side effect most of the time) of the particular characteristics of our bodies. Much the same applies to cultural knowledge, with the remarkable difference that quite often the 'virus' of culture is not harmful; instead, it can even be very useful to its hosts. We shall come back to this. According to the evolutionary history of the human brain that we have just seen, the ability to acquire cultural knowledge could not have been the selective advantage conferred by ToM. Rather, human capacity to produce and transmit cultural knowledge is merely a by-product of our ToM.

Given the fact that cultural knowledge is so important to us, one may wonder why I claim that ToM was not selected for transmitting cultural knowledge. We can answer this question with an exercise in logic. What could be the advantage of having a device that enables us to get cultural knowledge when there is no cultural knowledge to be acquired? Let us imagine the first person born with ToM mutation. By definition, cultural knowledge is a shared function. But when the first person with ToM mutation was born, there was no cultural knowledge to acquire, precisely because no other species members had ToM. By following this logic, we know that the acquisition of cultural knowledge could not be the reason why ToM prospered in human evolution; the acquisition of cultural knowledge was just a side effect of that device. The selective advantage that the ToM conferred on that first person was probably to improve cooperation with members of his or her species, to facilitate interaction with her offspring (if that mutant was a female), or to use tools made by other species' members. In fact, it is most likely that ToM conferred all three advantages at more or less at the same time. These are functions that most likely helped that mutant to outreproduce other members of his or her species. But none of these functions is directly related to the production of cultural knowledge – with the possible exception of the sharing of tools, because by sharing tools with someone else I am beginning to share the knowledge that goes into the production of that tool.

Whatever the case, the side effect that ToM gave rise to – the emergence of cultural knowledge – turned out to have absolutely revolutionary consequences for the history of life on this planet. Let us have a look at the main characteristics of cultural knowledge in comparison with the other forms of knowledge that humans share with other organisms (individual and genetic knowledge):

1. Cultural knowledge is not innate. We are not born with it. Instead, we acquire it; we learn it through interaction, not with the environment, as is the case with individual knowledge, but with other humans.
2. Unlike individual knowledge, and similar to genetic knowledge, cultural knowledge can be accumulated. It is always collective knowledge. It is shared by a group of individuals, but not by all the members of the human species. This is because in order to acquire cultural knowledge you have to interact with the individual who possesses that knowledge (either directly or indirectly – by interacting with an individual who interacted with an individual who interacted with that individual, etc.). Because our interactions with the members of our species are limited – we cannot interact with all the members of the human species that exist and that have existed so far, not even indirectly – our cultural knowledge is only shared with a limited group of individuals. We are beginning to see the reasons for cultural differences.
3. Cultural knowledge can be embodied not only in ideas, theories and thoughts shared through talking, but also in material things. Not only written texts, but also manufactured objects are embodiments of cultural knowledge. Think of all the manufactured objects that surround you at this very moment and of all the cultural knowledge that they embody, namely the cultural knowledge that goes into their production.
4. As in genetic knowledge, cultural knowledge does not need much cognitive processing. It can also become almost 'automatic' and unconscious. Remember that one of the shortcomings of individual knowledge is that it requires a cognitive investment: you need to be able to remember your experiences of your interactions with the environment and to process the information that comes from those experiences – or from the memories you have of those experiences. Even though this can also become unconscious, the difference between cultural and individual knowledge is still important. Cultural knowledge needs a bit of cognitive processing, too. You have to learn it from somebody and you need some special 'software' (our ToM) to do so. But that cognitive processing is very small compared with the huge amount of knowledge that you can activate in a single act of learning from another individual. Because all cultural knowledge is collective, any particular item of cultural knowledge is also always collective knowledge. But in order to acquire that cultural knowledge you do not need to process in your mind all the individual bits of knowledge that compose it. For this reason, cultural knowledge is also cognitively cheap

in comparative terms (especially compared to individual knowledge). It might not be as cheap as genetic knowledge, for which no cognitive processing is needed, but it is still very cheap given its (in evolutionary terms) absolutely fabulous results.

Just imagine the massive amount of knowledge that goes into the production of a computer. Nobody would be ever able to make even the simplest computer if he or she had to start from scratch with raw materials. It would be utterly impossible. In a way, it could be said that cultural knowledge has all the advantages of both individual and genetic knowledge without their disadvantages. As we have seen, genetic knowledge is very rigid because it can only change by means of random mutations and natural selection and at a very slow pace, which depends on the organism's reproductive cycle. Cultural knowledge is more flexible, although not as flexible as individual knowledge. Individual knowledge can change rapidly: I may think it is not raining and suddenly look outside and notice that it is indeed raining. My interaction with the environment has changed my knowledge of the weather in just a few seconds. Of course, individual knowledge can also give rise to individual customary behaviours that might not be easily altered. In any case, it is always easier to change an individual's habits, no matter how entrenched these happen to be, than the cultural norms of a whole society. Cultural knowledge can change as well, but in a much slower and more fragmentary way.

The cultural knowledge acquired by an individual changes throughout his or her lifetime. At the beginning of our lives we have very little cultural knowledge (and very little individual knowledge as well; we only have genetic knowledge). But then as we grow older we acquire more and more. We may be brought up in a particular culture, acquire a certain kind of cultural knowledge as children, and then emigrate as adults to live with people of a different culture. We interact with different people and therefore we learn different things in that new society. So the set of all our cultural knowledge might change as a result. But these are never radical changes. You cannot change your cultural knowledge 100 percent. If you did, you would fall into a state of utter confusion.

The cultural knowledge of a society (remember that cultural knowledge is always shared by a group of people) also changes through time. The cultural knowledge that we have right now is different from the cultural knowledge our ancestors in this very same society had, say, one hundred years ago, two hundred years ago or even just fifty years ago. In Chapter 4 we shall take a look at the possible rules that might determine cultural change, a very controversial issue as we will see. But cultural changes in societies through history are never radical changes, and when they are radical they can give rise to situations of confusion and anomie, which is the feeling of the absence of moral or social norms. This has occurred, for instance, in some colonised societies that in a relatively short time see most of their cultural system destroyed by an external colonial power. In any case, no matter how slow and partial cultural change is, it is always faster and more radical than changes in genetic knowledge. We have practically the same genome as humans who lived

100,000 years ago. Our genes are much the same, but our cultures have experienced radical changes.

The relatively fast pace of cultural change means that we can use our cultural knowledge to adapt to changing environments without changing our genetic knowledge. We can also do this with our individual knowledge (as can other animals that possess individual knowledge). But there is a very important difference with individual knowledge. Whereas individual knowledge is the knowledge gained by one individual throughout his or her life (and kept in his or her memory), cultural knowledge is collective knowledge: it is all the knowledge shared by the members of a society and produced not only by those who are living at a particular time, but also by their ancestors. The cultural knowledge of any society – even the simplest ones – is enormous compared with the individual knowledge of any human being. This is because cultural knowledge, like genetic knowledge, can be transmitted from generation to generation.

Cultural knowledge and adaptation

Cultural knowledge can be a very powerful mechanism of adaptation, much more powerful than individual knowledge (which is very limited), since it draws from the experiences of thousands of individuals. Cultural knowledge is also much more flexible than genetic knowledge, which is not as limited as individual knowledge but is very rigid. Thanks to our cultural knowledge, we humans have been able to live and reproduce in practically all environments on earth and dominate all other living organisms. And yet cultural knowledge is not perfect. It has at least one shortcoming, although you might say that it is a minor one, given all the advantages that it provides. I have just said that it is (or can be) an excellent mechanism of adaptation, i.e. to increase the biological fitness of the human species in practically any environment. The problem is, however, that cultural knowledge does not have to be *always* adaptive.

What does it mean for any form of knowledge to be non-adaptive or maladaptive? It simply means that that knowledge does not increase the biological fitness of the organism that has it. Individual knowledge can be maladaptive, for instance, when for whatever reason I obtain incorrect knowledge about the environment. I eat a poisonous plant, because I have mistaken it for an edible one. If I die as a result, or I fall ill or in some way damage my reproductive fitness, that maladaptive individual knowledge ('this plant is good to eat') will die with me, since I cannot transmit my individual knowledge to anyone else. Genetic knowledge can also be maladaptive when the environment where a particular species lives changes. When a change in environmental conditions takes place, that particular species will not be able to reproduce in this new environment, or will not reproduce as much as another one that is better adapted. Therefore its genetic knowledge will turn out to be maladaptive and that maladaptive genetic knowledge will eventually disappear, either because the species goes extinct or a mutation takes places that eliminates or disables those maladaptive genes. At a different scale, that is precisely what happens when a maladaptive mutation crops up in a particular individual. That individual, by definition, will not reproduce as much as individuals of the species who do not carry the mutation. Consequently, the mutation will not replicate and will instead disappear.

But the situation with cultural knowledge is entirely different. A non-adaptive item of cultural knowledge is, again, that which hinders the reproductive fitness of its possessor. But cultural knowledge can reproduce itself irrespective of the biological reproduction of its possessor. This is so because cultural knowledge and genetic knowledge may circulate on different tracks. I can only obtain genes from my parents; no one gets genes from a person who has not managed to reproduce. Genes only circulate vertically. (True, some organisms seem to be capable of transmitting genes horizontally, but that is an exception to the rule that should not detain us now.) Many of the cultural messages that one acquires also come from one's parents. But there are lots of other cultural messages that may come from other sources. Cultural messages circulate vertically, horizontally and, most importantly, obliquely. Notably, I can get cultural messages from people who have not reproduced. Thus, cultural knowledge that turns out to be maladaptive in the sense that it prevents people from reproducing, or makes them reproduce at a lower rate than those who have not received those cultural messages, can nevertheless spread. Here is where the analogy between culture and viruses of the mind turns out to be particularly insightful. A virus might kill you, like the flu virus or the HIV virus. But viruses continue to exist because before killing their hosts they have managed to colonise another body – otherwise they would die with their hosts and that would be the end of them both. The very same thing happens with cultural messages. A cultural message might kill me, or perhaps just prevent me from reproducing as much as someone who has not got that message. And yet before I die there are lots of ways that message can spread to other minds, which can be minds other than those of my children (if I have any).

To sum up the argument we have developed so far, it could be said that cultural knowledge can relate to the human mind in three different ways:

1. Symbiotic relationship. This is adaptive cultural knowledge, that is, cultural knowledge that increases the reproductive fitness of those who acquire it. For instance, think of the cultural knowledge that enables an Inuit hunter to survive in the polar region.
2. Commensalistic relationship. This is cultural knowledge that makes use of the human mind without increasing or decreasing its reproductive fitness. Its effects are neutral, neither adaptive nor maladaptive. Think of university education. Will university graduates have more children than those without any university degree? In most countries university graduates are not particularly prolific or more prolific than other people.
3. Parasitic relationship. This is maladaptive cultural knowledge, which is cultural knowledge that hampers the reproductive fitness of those who acquire it. Think of contraceptive practices: these are cultural practices that make use of

our sexual instincts but prevent us from reproducing. Notice that maladaptive cultural knowledge does not have to be 'bad' in any moral sense. We might think that it is good for us, for our society or for our economy not to have too many children. And yet, biologically speaking, our behaviour turns out to be maladaptive for the very simple reason that we are having fewer children because of it.

This is a paradox of culture. The mechanism that enables humans to adapt so well to so many different environments is, at the same time, a mechanism that produces non-adaptive messages. All human cultures have adaptive and non-adaptive traits, because, as some argue, this is the price we have to pay to enjoy all the benefits that cultural knowledge provides.

Free-riders of the brain

So if it does not have to be adaptive, what characteristics must cultural knowledge have in order to be able to colonise human minds? It must find the appropriate conditions for reproduction. Cultural knowledge piggybacks on whatever structures it can find in the mind and makes use of them for its own 'purposes', that is, to reproduce itself by colonising another mind. Needless to say, this is not meant to attribute agency to cultural knowledge. In the same way as there are not 'selfish' genes, nor does 'Mother Nature' design anything; cultural knowledge has no actual 'purpose'. It is the magic of natural selection that makes everything look as if biological or cultural entities followed their own intentions. But in fact they have none.

When the structures it uses are part of our genetic knowledge, we say that cultural knowledge free-rides on our instincts. If they are part of our individual knowledge, we say that cultural knowledge free-rides on our experience. And when those structures are part of our previous cultural knowledge, we say that new cultural knowledge free-rides on pre-existing cultural knowledge. Think of the messages that at this very minute are entering your mind while you are reading this text. What conditions will enable them to replicate? First of all, some biological conditions must be met. You need some kind of brain to understand these messages, and it has to be a human brain (genetic knowledge). But this brain must already have some cultural knowledge in it so you can understand what you are reading: it must have reading skills, knowledge of English, some previous form of education, etc. (cultural knowledge). Furthermore, different aspects of your individual experience might make you more susceptible to understanding and liking these messages (individual knowledge). All this is what makes your mind liable to become the host of the cultural messages you are receiving right now.

A few examples will make this clearer. The first one is the Hindu caste system. Traditional society in India is divided into a set of hierarchically-organised endogamous groups called castes. ('Endogamous' means that any one person can only marry someone belonging to his or her own caste.) These groups constituted the backbone of Hindu society and all the major political and economic functions were conferred upon them. There were very rigid norms regulating interaction between members of different castes based upon the notion of purity and pollution. Superior castes were seen as purer than inferior ones and, among other things, their members could not touch food that had been in contact with a member of an inferior caste. But how could this ideology of purity and pollution as regards human beings have been so pervasive in Hindu history? For those of us who have not been brought up in this particular culture, we might find it hard to believe that anyone could see another human being as a source of pollution just because he or she belongs to an inferior caste. But the longevity of this cultural ideology is not due to any adaptive function in Hindu society. Avoiding food that had been touched by the member of a lower caste does not increase your reproductive fitness in any way. Caste prejudices survived because they took advantage of important instinctive reactions rooted in our immune system. All humans are particularly fussy about the food they eat since food can be a source of dangerous germs and contaminants. But our hard-wired wariness about food can be substantially shaped by cultural education. We all find dirty food disgusting and, as a result, we are very sensitive to messages that alert us to the characteristics of food. Persistent information concerning the 'polluted' nature of a particular food item – because it has been touched by some undesirable person – will very easily penetrate our minds and make us react accordingly. Westerners do not find anything disgusting about food touched by a particular person, but this is just because they have not been brought up in that culture.

A second example comes from what we could define as 'sweet culture'. Many human societies, including Western societies, have a particularly powerful sweets industry. A great amount of money is spent every day in advertising them and on buying them. Practically all societies have hundreds of recipes for preparing sweet meals. Is this sweet culture adaptive in any way? Probably not. We put on weight if we eat too many sweet things and are prone to suffer from heart disease. Clearly, our sweet tooth (which we share with other mammals) had an adaptive function in ancestral times. Sweet stuff is rich in calories, which enable our bodies to metabolise huge quantities of energy in a relatively short time. Desire for and consumption of sweet food is adaptive in a context of constant physical exercise and, specifically, in an environment where sweet foods are not particularly abundant, such as any natural environment. But we humans, especially in Western societies, have changed that environment thanks to our cultural knowledge that has created a powerful food industry. All the cultural practices related to the consumption of sweet foods thrive, not because they are adaptive (they are not), but because they free-ride on our innate fondness of sweet stuff.

A third and final example is provided by the belief in spirits. This is one of the most widespread beliefs in all human religions: the idea that there are invisible beings around us, normally rather powerful, some of them friendly but many of them malevolent, capable of causing all sorts of mischiefs. Each culture might define spirits in a particular way: gods, elves, fairies, leprechauns, demons, sprites, ghosts or spirits of the dead. But they all have one characteristic in common: they are normally invisible, though sometimes they can manifest themselves in some (usually scary) way. How is it that humans in so many different societies with different cultures end up believing in these invisible beings? What is the point of believing in something that you can't see? Primates in general are very vulnerable in the dark. Unlike other mammals, we have a relatively weak sense of smell and a relatively strong sense of sight. Therefore, when we are in the dark we cannot receive much information from the environment, other than auditory information. In these conditions, the littlest noise is likely to put us on the alert. What is that? Who is there? Our nervous system is automatically ready to face some unknown danger. That was clearly an adaptive reaction for our ancestors, since only those who were on the alert and ready to act in that situation were likely to survive the attacks of predators or enemies who might be lurking in the dark. That is why we all become somewhat restless when we hear noises and can't see what produces them. (Note how important the soundtrack is in terror movies.) So the cultural belief in invisible beings, beings that might come after us, we cannot see them, but they are there, if we listen carefully maybe there is some noise that gives them away... this belief sticks very easily in our minds, for all that it can be perfectly maladaptive.

But not all cultural knowledge is like that. I have already pointed out that thanks to our cultural knowledge, humans can live in practically any environment on earth, and they have come to dominate all other species of living organisms. So cultures can certainly be adaptive. But my purpose in offering these examples of maladaptive cultural knowledge was to highlight a very important characteristic of this kind of knowledge: the existence of cultural knowledge is not related to whether or not it is biologically adaptive for human beings. Adaptation is all-pervasive in the biological world. That is how biologists explain the characteristics of any one species. Maladaptations crop up from time to time, especially in changing environments. But because they are faced with continuing natural selection, none of them survive for very long. Nothing of that sort exists in the human cultural world. Surely, very maladaptive cultural practices may lead to the utter extinction of those humans who adopt them, and there is no way cultures can survive without humans and their minds. But most maladaptive cultural messages can spread well before the biological extinction of their human hosts. That could be one of the reasons why many social scientists customarily thought that biology had nothing to do with culture. According to this point of view, evolution by natural selection applies to natural history but not to the cultural history of human societies, so we can handsomely disregard it in the study of culture. But this view is wrong. It is wrong because no matter how maladaptive particular cultural practices happen to be, they still have to colonise human brains in order to survive. And all human brains start off as the products of the biological evolution of our species.

Counterintuitive culture

Maladaptive cultural messages are those which most closely resemble a virus: a harmful guest that takes advantage of its hosts' capacity to interact with other individuals in order to produce copies of itself at the expense of its hosts. Now, apart from being maladaptive, a cultural message can also be counterintuitive. This is a very important concept and will be specifically relevant to what we will see in Chapter 4, but I will raise the issue here briefly.

Compare the following two cultural practices: contraceptive sex and compulsory celibacy. Both of them can be said to be maladaptive in that they can prevent individuals from reproducing. (For the purposes of this example, we shall leave aside the case of family planning, in which contraception is indeed adaptive because it tries to match the number and timing of offspring to the available resources, including the desire to parent.) But only the second practice is counterintuitive in addition to being maladaptive. Contraceptive sex can have conspicuous maladaptive consequences but it does not contravene our intuitions in any way. Far from it, contraceptive sex can be seen as an example of maladaptive cultural knowledge that thrives by colonising our sexual instincts. So in a way contraceptive sex simultaneously coincides with and contradicts our genetic knowledge. It coincides with our genetic knowledge in that it is perfectly consistent with our generalised desire to have sex, which is how natural selection enforces reproductive behaviour upon all sexually reproducing species. But it also contradicts our genetic knowledge to the extent that it eventually removes the reproductive results of sexual behaviour. Consequently, we define it as maladaptive. Now compulsory celibacy, by contrast, in addition to being equally maladaptive, also clashes with our sexual instincts. That's why we call it counterintuitive.

Note that by drawing this distinction between the maladaptive and the counterintuitive, we can conceive not only of intuitive and maladaptive cultural practices, such as contraceptive sex or the counterintuitive and maladaptive, such as compulsory celibacy, but also of those that are counterintuitive and adaptive. A clear instance of this would be diet cultures. In a context characterised by a superabundance of sweet foods, the counterintuitive message that discourages us from indulging in our sweet tooth is both adaptive and counterintuitive. Clearly, adaptive counterintuitive cultural messages presuppose some sort of biological maladaptation, normally because our culturally constructed environment turns out to be incompatible with our environment of evolutionary adaptedness (EEA).

The important thing to keep in mind about both adaptive and maladaptive counterintuitive cultural knowledge is that special means are needed for its assimilation. This is a form of 'unnatural' knowledge, so we need something more than just normal communication (see next section) for it to be successfully transmitted. People have to be trained to assimilate counterintuitive cultural messages, and the more counterintuitive they are, the more rigorous the training required. Nobody would accept a life of compulsory celibacy without some sort of contrived cultural indoctrination. And the same applies to stringent diets, especially if they are accompanied by some form of physical exercise, which we have to follow if we are to avoid the deleterious consequences of our modern food industry and lack of physical activity. Or think of one of the most counterintuitive forms of cultural knowledge that human societies have ever produced, and a very successful form of knowledge at that, namely, scientific knowledge. Long years of very special training are normally needed for anyone willing to master the intricacies of scientific postulates and theorems. But how can human societies be persuaded that these counterintuitive forms of cultural knowledge are 'good' for their members (namely, adaptive) in any way? Do they have to be 'aware' of this evolutionary logic? Again, we will have to wait until Chapter 4 for a full treatment of this intriguing issue.

The transmission of cultural knowledge

Non-adaptive genetic knowledge ends up disappearing because, by definition, it hinders the reproductive fitness of the organisms that have this knowledge, and genetic knowledge can only be transmitted through biological reproduction. This is not a problem for individual knowledge since, by definition, individual knowledge cannot be transmitted. Both adaptive and non-adaptive individual forms of knowledge disappear with the individuals that carry them. The situation with cultural knowledge is entirely different. Both adaptive and non-adaptive forms of cultural knowledge can be equally transmitted, because the capacity of cultural knowledge to replicate itself is not the result of its adaptive value. So how is cultural knowledge transmitted from one individual to another? This is a fundamental question because, according to the epidemiological model of cultural reproduction that I espouse, the very existence of cultural messages depends on their ability to produce copies of themselves.

Culture and language

We assimilate cultural knowledge through communication. In this section we will be looking at one of the most important means humans use to communicate cultural knowledge, namely, language. Language and culture are so closely interrelated that very often we tend to confuse them. Anyone who has a language has a culture and vice versa. But there is not a one-to-one correspondence between language and culture. There can be two (or more) different cultural groups that use a single language and likewise there can be two (or more) different languages whose speakers share one overarching culture. Moreover, I can use one language to describe the content of a totally unrelated culture. This is in part because the link between sound and meaning in a language is arbitrary; there is no particular reason why each word has the meaning that it does. This means that language and culture are not equivalent. A language is merely an instrument of communication. Some prefer to define the main function of language as being an instrument of thought rather than communication. In any case, my point is that language does not have any meaning in itself – unlike cultures, which we have defined as 'systems of meanings' – other than, perhaps, the need or will to think or to communicate. There might be some additional meanings attached to particular languages in certain historical circumstances: for instance, some populations manifest their national or ethnic identity by using their language. But that is purely an accidental meaning attached to that language. It has nothing to do with the essential function of language, which is to make human communication and thought possible.

The second confusion about human languages has to do with their origins. It was normally assumed that languages are learnt, for we are born without a language and we end up speaking the language of the people around us. So it seemed as if language was part of our cultural knowledge and we learnt a language the same way as we learnt the rest of our cultural knowledge. But that confusion was eventually dispelled by American linguist Noam Chomsky in the 1960s. We have seen this already when we were talking about cognitive modules and the LAD, so no need to repeat it now. The problem, however, as I pointed out then, is that, if we accept Chomsky's theory of the innateness of our capacity to learn a language, we still have to explain its evolutionary origins. What sort of selective pressures acted upon our mute ancestors to make the capacity to learn a language adaptive in that environment? Naturally, that capacity only makes sense once languages already exist. But it is pointless to have the capacity to learn a language when there is no language to learn!

We had a similar problem when we were trying to account for the evolutionary origins of our capacity to assimilate cultural knowledge. We concluded then that our capacity to learn a culture emerges from our ToM module. But that module had not been designed by natural selection for that purpose. That is why the definition of culture as a 'virus' of the brain is so persuasive. The late American biologist Stephen Jay Gould put forward the concept of 'exaptation' or 'spandrel' to refer to this phenomenon (1991). According to Gould, exaptations or spandrels are very common in evolution. A particular biological trait that has been selected because it confers a particular advantage ends up serving a different function. He gave the example of the wings of insects and the feathers of birds. It seems that originally they were selected for thermal regulation. But eventually they were used for flying. So flying is not the biological 'function' of insects wings (or it was not the original function); therefore we could say that wings are 'exapted' for flying, not adapted. In another paper, Gould used the concept of 'spandrel' to refer to the same thing. In an article entitled precisely 'The spandrels of San Marco' (1979), Gould and the American geneticist Richard Lewontin argued that some biological characteristics of an organism are not adaptations but mere by-products of something else. And they used as an illustration the metaphor of the spandrels of St Mark's cathedral in Venice. Spandrels are the adjoining spaces between two arches the existence of which is merely a by-product of the way the arches are built. In San Marco, and many other places, however, they are painted with scenes from the Bible. But this is clearly not their function; their function is merely to solve an architectonic problem: how to hold the dome by means of these arches.

The same thing happens with many biological traits, a clear example of which would be our ToM. Initially it might have been selected in order to facilitate communication and coordination of large cooperative groups, and eventually it was used to accommodate the virus of culture. We do not know what the initial function of our LAD (or some other biological trait that can be seen as the precursor of our

LAD) could have been, before it was 'co-opted' by language. It could be argued that once some form of protolanguage is in place, a greater capacity to acquire and use that language could improve communication between any two individuals who happen to share that capacity, say, a mother and her child. A mutant female with that enhanced capacity would enjoy a selective advantage in relation to the other females for she could better communicate with her children, who are also likely to have that mutant gene for enhanced linguistic abilities. Notice that this does not equally apply to males for they are never certain of their paternity – though they might be in a monogamous community. That might explain why women the world over are not only better at figuring out other people's mental states but also better talkers.

Whatever the case, we already know that only humans can learn a language. But non-human animals also seem to be able to communicate among themselves (birds, whales, etc.). What could be the difference between human languages and these animal languages? This is a very controversial issue. We know that human languages are qualitatively different from the languages used by non-human animals. But we are still not sure what the nature of this qualitative difference is. Human languages are an evolutionary mystery: there are no precedents of human languages in the animal world, no intermediate stages between human languages and non-human languages. Some idiosyncratic characteristics of human languages have been posited as likely candidates for the key difference between human and non-human systems of communication. According to one school of thought, syntax is what animal languages do not have. For instance, vervet monkeys have been recorded producing different types of alarm calls, depending on the predator that is approaching. But they cannot combine different howls in order to refer to something different. And that is, presumably, because they do not have syntax.

Other authors think that the crucial difference has to do with so-called symbolic reference. Human languages are symbolic; non-human languages are not. That means that the relationship between words and the things they refer to is totally arbitrary; it is conventional. This enables humans to invent new words to refer to new things or new experiences. Animals cannot do this. A symbolic language enables me to disentangle words from objects: I can pronounce words without seeing the objects they refer to. For instance, I can talk about the car I bought yesterday. My words have meaning even though you cannot see the car. Maybe I am lying to you and the car does not even exist. And yet you can understand what I am saying. Conversely, I cannot laugh or cry without the corresponding emotion (unless I am pretending, but that is always more difficult than simply telling lies). And vervet monkeys cannot produce calls outside the situations that trigger them. The relationship between the vervet monkey's call and what it stands for would not be symbolic but indexical, following Pierce's well-known definition. The call cannot be separated from its object because the call is caused by its object, in the same way as a laugh or a cry are caused by the corresponding emotion or the rain is caused by the clouds.

The two things are undoubtedly related. Symbolic reference is possible because words are not related to things in the first place but to other words, and syntax is precisely what enables human brains to relate words to other words. But perhaps the most important difference between human and non-human languages is that humans use language to refer to our intentions, to make them manifest and to grasp other people's intentions. Cognitive scientist Thom Scott-Phillips has called the communication of intentions ostensive or ostensive-inferential communication, by contrast with the communication of information, which he terms code model communication (see bibliographical note). Animals cannot communicate mental states because they do not have a ToM. That probably explains why their languages are so different from ours, because their function is not to communicate mental states. Their function is to communicate information. They only have code model communication, in Scott-Phillips's terms. For instance, females of certain species give off a certain smell when they are in oestrous that arouses males and incites them to copulate with them. We could define this smell as a 'language' of sorts that conveys information, like 'I am in oestrous; I am ready to copulate'. But this is only a metaphor, since neither does the female have any intention of communicating anything nor is the male grasping any intention. It is, we might as well say, an automatic process: the female is in oestrous and gives off that characteristic smell; the male senses the smell and copulates with the female. The same thing can be observed between certain artefacts such as thermostats and heating systems. A thermostat senses a change in the temperature and turns the heating system on and off accordingly. But neither the thermostat nor the heating system has intentions; they do not have mental states to communicate to each other because they do not have minds. Again, the relationship between changes in temperature and the turning of the heating system on or off, as the case may be, is causal, not symbolic or conventional and the same applies to the smell of the female and the reaction of the male.

True, I can infer some very simple mental states, such as basic emotions like fear, hunger etc. without explicit linguistic communication. This is why I can attribute intentions to non-human animals, for instance, even rudimentary 'beliefs' (the dog believes that his master is coming) or 'desires' (the cat is hungry). But without language I would not be able to communicate those intentions to someone else. Neither the dog nor the cat wants us to know what they are up to; it is we who 'infer' those mental states from their behaviour. Note the important difference with human linguistic communication: the dog might want us to get food for him, but it is the food that he wants, not us knowing that he wants food. Pre-verbal children, on the other hand, can easily understand other people's intentions and communicate theirs, provided that they are simple intentions (such as 'I want this', 'I don't like that'). So in this case we do certainly have communication of intentions without language. But they can neither communicate nor understand complex states of mind, such as false beliefs, for instance, in the absence of language.

We humans have 'remnants' of non-human animal languages. For instance, sometimes when we are very angry we start to shout at each other. Why do we do that? Probably because we are trying to scare the other person with our shouting. But sometimes we might do it instinctively, automatically, as if it was not our mind but our body that 'decided' to shout, the same as dogs that bark when they see danger coming. Note the difference with linguistic communication. Both the angry shouting and the fear it instils into someone else are instinctive reactions, whereas there is nothing instinctive in linguistic communication. By instinctive reactions I mean behaviours hard-wired in the nervous system of the animal, which has a particular evolutionary history. By barking and showing their teeth to possible attackers, dogs, or their ancestors, probably increased their capacity for survival and reproduction. Nothing of that exists in linguistic communication. In fact, different parts of the brain seem to be involved in linguistic and non-linguistic communication: prefrontal cortex for the former and cerebellum for the latter.

So human languages are used to convey the speaker's intentions, that is to say, to make our minds communicate with each other. And that is why the relationship between language and cultural knowledge is so close. We can sum up the relationships between language and culture with two main principles. First, without language we would not be able to communicate the contents of our minds to other people; consequently, even though we still might have a ToM, cultural knowledge could not exist. Second, without a ToM, language could not serve any function, since we would not have any mental states to communicate.

Culture beyond language

Language is the instrument of culture almost by definition. This does not mean, however, that we always assimilate cultural knowledge by means of explicit verbal communication. Think of any ordinary conversation. We normally do not make explicit everything we want to say because we take it for granted that the other person already knows part of it. Sometimes it is quite difficult to understand other people's conversations because we do not know what they are about; we do not share the implicit knowledge that each party assumes the other must have. This is called the indexicality of human communication, which means that in addition to their explicit, referential meaning, our words also generate meaning by 'pointing' to the implicit knowledge that we share with our interlocutors. In this sense, whatever we say always leaves something 'unsaid'. The less implicit knowledge we share with the person we are talking with, the more explicit needs to be our verbal communication.

Think of the way children learn from adults, the way they assimilate the cultural knowledge of the society where they are being brought up. At times it may be through explicit verbal instructions, like when parents tell them what to do and what not to do or when teachers give them lessons. But quite often they learn by imitation: they observe adults behaving or speaking in a particular way and they mimic what they see. Notice as well that these non-linguistic forms of culture learning normally take place in language-saturated environments. In other words, even though lots of cultural information enters our minds without explicit verbal instruction, linguistic communication acts as an implicit scaffolding to make this non-linguistic communication possible. Countless cultural messages enter our minds in the form of habits, practices, lifestyles and ways of doing or not doing things. We are not even aware that we have them, and that is because we were hardly aware that they were getting into our minds in the first place. But they constitute an unconscious bedrock upon which more conscious cultural forms can be built, sometimes in plain contradiction with what remains unconscious – the 'unsaid' of human communication.

We can imagine the human mind as a sort of basket that is being filled up with different layers of objects. At any one time, we can only see the last layer of objects; the others remain hidden underneath, even though they are still there. The same thing happens with our minds. We are only conscious of the last 'layer' of cultural messages to enter our minds. As more messages are added, the older messages are 'covered up' and we are no longer aware of them.

With so much unconscious cultural knowledge stored in our minds, it is no wonder that explicit verbal statements concerning what we think may contradict the content of our cultural unconscious. Anthropologists Claudia Strauss and Naomi Quinn gave the example of gender relationships. We learn a lot of our gendered behaviour through non-verbal instruction; little girls imitate their mothers and little boys their fathers. The sort of gendered cultural information that we might learn in this way might be very unbalanced, for instance, and 'sexist'. But this escapes our awareness. Once we grow up, we receive explicit messages about gender equality and against sex discrimination. Even though we are likely to consciously accept theories of gender equality, our unconscious sexist habits might not change as a result.

The problem of meaning

We already know how culture enters our minds, as a virus that colonises our brains. Now we will explore what happens afterwards; that is, how culture determines human behaviour. How does culture exit the mind it has colonised in order to colonise other minds? As we have seen, that is one of the ways in which cultural messages replicate themselves. Suppose I receive a cultural message that tells me how to behave in a certain situation, say, by observing other people's behaviour. That message can replicate itself in two ways: either I teach other people how to behave in that particular situation, through explicit verbal communication, or I myself behave in that way so that other people can learn just by looking at me (which happens to be the way I myself learnt the behaviour). Recall that cultural knowledge exists as long as it can replicate itself. A cultural message that fully colonises my mind but stops there without going out to colonise any other mind is a dead message. Again, the parallel between cultural messages and viruses is very informative.

Human behaviour appears to be rather more complex than the behaviour of non-human animals. Whereas for non-human animals we only need to take into account their genetic and individual forms of knowledge, in humans we need to take into account cultural knowledge as well. But how can we establish a connection between a particular form of behaviour and a particular piece of cultural knowledge? How does culture determine or act upon human behaviour?

Culture and mind as emergent properties

So far we have been talking about mind and culture in a rather loose way. We have compared cultural knowledge to genetic and individual knowledge. We have also seen that cultural knowledge originates in the interaction of minds that have to be endowed with ToM, a special property that enables those minds to figure out what each other has inside. But what is exactly the mind that our ToM enables us to figure out? What we find in each other's minds are mental representations: thoughts, intentions, desires, beliefs. But what is a mental representation? What is a thought? And what has all this to do with the brain, if anything? We know that a brain is a gelatinous spongy object inside our skulls. It is made up of very small entities we call neurons that we can see with a powerful microscope. But what about our thoughts? Our brain produces our thoughts, or so it seems. But where are they exactly? *What* are they? The accumulation of thoughts resulting from the interaction of human minds is what we know as 'culture'. But as physical objects neither minds nor cultures are anywhere to be seen: invisible things producing more invisible things. We seem to be on pretty shaky ground.

Imagine a traffic jam of hundreds of cars stuck on the road. They go very slowly, stopping again and again. Nothing in the characteristics of each particular car explains why they move that way. Each one could travel much faster; their engines are in perfect condition. But they can't go any faster because of the traffic jam. But what is the traffic jam? It is simply the effect of the accumulation of lots of cars in a small space. In strictly physical terms, it is not a new entity; there are only cars on the road. The traffic jam is an *emergent property* of the interaction between cars. We call it an emergent property because the traffic jam has characteristics that are different from the characteristics of its component objects, the individual cars. Not only that, emergent properties are likely to constrain the behaviour of their component objects. This is called 'top-down' causation. Cars have produced the traffic jam, but the traffic jam determines their movement. That sounds paradoxical but it is not really. We are actually surrounded by emergent properties of different kinds. Look at the letters of the very same text you are reading right now. They are simply marks of ink on a sheet of paper (or sets of pixels on a screen, if you are reading a digital version), but when they are put together in a certain order they produce another emergent property, something different from each individual letter and that determines the ways in which they can be arranged. Sets of letters become words and sets of words become sentences. Again, neither words nor sentences are nothing other than sets of letters. And yet, words and sentences constrain the letters that can be written and in which order.

Accordingly, an emergent property is something that originates out of the interaction of a set of pre-existing entities, and once it has emerged it will constrain the behaviour of those pre-existing entities. Let us think of the mind as an emergent property of a set of interacting neurons. In strictly physical terms, there is nothing but the neurons and their interactions. But out of these interactions emerges a new property that will constrain the behaviour of those neurons, in pretty much the same way as the traffic jam constrains the movements of the cars. This would be parallel to the way that words and sentences determine which sounds I produce and in which order. But words and sentences are nothing but sets of sounds arranged in a particular order (sounds 'in interaction' with each other). Now turning to the human brain, we can use these examples to understand emergence as an answer to the so-called 'mind-body problem': how can the mind, an immaterial entity, move the body, a material entity? I want to raise my hand and, lo and behold, my hand rises. It looks as if the intention had activated the appropriate neural connections and turned on my nervous system, tensed the muscles of my arm and eventually raised my hand. But in fact, desires and intentions are nothing but sets of neurons in interaction with each other, just as a traffic jam is a set of cars in interaction with each other, and a word is a set of sounds in interaction with each other.

Now let us think of culture as an emergent property of a set of interacting minds. Not any kind of mind or set of minds can produce a culture; they have to be human minds, endowed with the capacity to interact with each other thanks to their ToM module. A culture is, as we shall see in a minute, a system of meanings. And a meaning is an intention as figured out by somebody else's mind. 'I know what you mean', signifies 'I know your intentions; I know what is in your mind'. In the same way as an intention determines the movements of my body and turns them into *behaviour*, a system of meanings is also likely to determine the desires and intentions inside my mind and therefore the behaviour produced by those desires and intentions, which will become *meaningful* behaviours. True, sometimes my body produces 'unintentional' movements, such as a twitch that is not the result of any intention generated in my mind. A nervous tic of this sort is not only unintended, it is also pointless. But I can also have unintended behaviours with a purpose, such as when I instinctively raise my arm to protect my eyes when I see a projectile coming toward me. Note that in this case this behaviour, though unintended, has a purpose. It is not 'my' purpose, because it hasn't been produced by my mind (or my intentions). Rather, we could say that it is my genes' purpose, because my genetic knowledge has produced it irrespective of my mind (that is, my individual knowledge). In short, I do not need to have had the experience of being hit by a projectile to act in this way. Similarly, I might have intentions underdetermined by any system of meanings, and therefore bound to produce meaningless behaviours. As I approach my car, I put my hand in my pocket to reach for the keys. Even though I do this without any communicative intention, it is a meaningful intentional behaviour and anyone who sees me doing it can easily figure out my intentions. Imagine that instead of taking out my keys I keep putting my hand in my pocket and taking it out with no apparent purpose. That would be an intentional but meaningless behaviour because nobody could understand why I perform it. We can certainly engage in meaningless behaviours, but if they were the general rule social life would be impossible. Minds could not interact with each other and cultures would not have emerged.

We shall see now in a bit more detail the way in which culture as a system of meanings determines human behaviour. This determination is very different indeed from the sort of determination that genetic knowledge ('genes') and individual knowledge ('minds') exert on behaviour.

Let us look first at genes. How do genes determine human behaviour, or the behaviour or any other animal? Genetic knowledge is what we have called instinctual behaviour. Genes are just DNA sequences that, by themselves or in interaction with other DNA sequences, produce chemicals called proteins. By means of complex chemical reactions in the body and the nervous system, these proteins make the animal behave in a particular way. If the light in this room gets too bright you instinctively close your eyes. You don't even have to process that information in your brain or think about what is going on. Your nervous system reacts automatically to the excess of light by closing your eyes.

Instinctual behaviours are often slightly more complex than this one. Think of the opposite situation, what happens when we enter a room where there is no light. In the dark, we tend to be more alert than where there is light. Our senses (specially our hearing) automatically become more awake, attentive and sensitive. We saw this when we were talking about the belief in spirits. Small children are instinctively, that is, automatically, afraid of the dark. Even very small children who have never had a bad experience in the dark tend to be afraid of darkness. Other animals are also afraid of darkness, especially primates, because of their poor sense of smell (as explained in Chapter 1). Fear of darkness in primates is therefore instinctive rather than originating in any individual experience. The reason for this fear (and its concomitant state of alertness) is that we primates are very vulnerable in the dark; so in ancestral times (probably when the first primates came into existence, that is, 65 million years ago), those who were afraid survived and passed on their genes, whereas those who were not afraid did not survive.

Notice, however, that not everybody is afraid of the dark in the same way, and not everybody reacts in exactly the same way. Instinctive behaviours often do not seem very accurate. This is why the typical metaphor of a computer programme can be misleading. Our nervous system, and the genome that produces it, is not a computer programme; that would be too costly in metabolic terms. For a particular gene to replicate itself it is not necessary for every individual with a copy of that gene to behave in exactly the same way under the same circumstances; it is only necessary for the *majority* to react in that way. Thus, if under certain stimuli the majority of animals with a particular gene behave in a particular way that happens to be adaptive, that gene will reproduce itself. This occurs even if there is a minority that, for whatever reason, do not behave in that way. As British anthropologist Robin Dunbar has put it, evolution is a statistical process, not a deterministic one.

So this is how (described in a somewhat simplified fashion) our genetic knowledge determines our behaviour. What about individual knowledge? How does it influence our behaviour? Remember that the mind is an emergent property of interacting neurons; therefore, individual knowledge determines our behaviour in the same way that a traffic jam determines the movements of the cars. In a way, both when our behaviour is determined by our genetic knowledge and when it is determined by our individual knowledge, there are only neurons acting upon our nervous system and producing particular movements. But from another point of view, as we have already pointed out, with individual knowledge there is something else going on here. Individual knowledge involves the emergent property we have called 'mind' that causes this new form of behaviour, which by definition becomes intentional behaviour.

An intended behaviour – that resulting from our individual knowledge – is produced by our cognitive system. Whereas the behaviour generated by genetic knowledge is 'automatic' for a certain stimulus, this is not the case for individual knowledge. For individual knowledge, information coming from the environment needs to be compared with previous experiences stored in the animal's memory. Depending on those previous experiences, the information coming from the environment will generate one behaviour or another. Several experiments can demonstrate this. For instance, you can teach a mouse to expect food when it hears a particular noise. The first time you make that noise and give food to the mouse, the animal probably will not 'understand' that the noise is announcing the arrival of food. But if you repeat the experiment a few times, the animal will learn that the noise is announcing food, so as soon as you make the noise it will expect the food. You can do this sort of experiment will all mammals (the famous Russian physiologist Ivan Pavlov used dogs), and with some birds as well. Note that after training, the behaviour of the animal produced by its individual knowledge becomes almost as predictable, as 'automatic' and unconscious, as its instinctual behaviour. But that's because the training process has introduced into the animal's mind such a powerful experience that it practically annuls any other experience that might be relevant in determining the animal's behaviour under those particular circumstances. But if we hadn't trained the animal's individual behaviour, its reaction would not be as easy to predict, because we would not know the contents of its individual knowledge, namely the myriad particular experiences that had shaped its mind throughout its life.

But what about culture? How does culture determine behaviour? Let us begin with the definition of culture we already know: culture is a system of meanings and symbols in terms of which humans govern their behaviour. To unpack this definition, we shall forget momentarily the word 'symbols' and concentrate our attention on the word 'meaning'. What is a 'meaning'? I argued above that a meaning is the other side of the coin of an 'intention'. When we say 'I understand what you mean', 'I know what you mean', or, conversely, 'I don't know what you mean', we are saying 'I understand your intentions'. And we grasp other people's intentions through our ToM. Remember that we have defined culture as an emergent property of interacting minds. What happens when minds keep interacting with each other on a regular basis for an extended period of time? When we define culture as a system of meanings we are saying that culture is a 'system of intentions'. All cultures are systems of the 'congealed' intentions of all of the people that have been brought up in a particular culture. That is, of all the people that have received a particular set of messages and have transmitted those messages to other people. It is important to emphasise that in this process of reception and transmission of cultural messages, our minds do not work as fax machines, merely replicating exactly the messages that they have received. Here is where the 'virus' metaphor of cultural transmission can be somewhat deceptive. We all receive lots of cultural messages from lots of different sources, and not everybody within the same society receives exactly the same messages, because not everybody has the same social interactions throughout his or her life. Furthermore, cultural knowledge gets mixed with the individual and genetic knowledge of each person and can never be replicated in exactly the same way in any two individuals, no matter how culturally (and genetically and experientially) similar those two individuals are.

Think of the culture of a particular society as a huge painting on which each person makes a small brushstroke. The final painting 'congeals' the intentions of all those who have made those brushstrokes, but none of them can be seen as the 'author' of that final painting. Needless to say, in any particular society, not everybody contributes the same number of brushstrokes. All societies have their remarkable cultural authors or cultural geniuses, particular individuals who can be said to be the authors of particular cultural objects more than anyone else: important writers, artists and scientists. In our own society, they are undoubtedly the authors of a lot more brushstrokes on the canvas of our cultural knowledge than ordinary people like you and me. But not even these cultural geniuses could have produced their works in isolation.

We know that cultural objects embody the intentions of their makers. This applies not only to manufactured objects, but also to the ideas, beliefs, theories and traditions that we learn when we grow up in a particular culture. These cultural objects embody the intentions of all those who have transmitted them. This is how we learn any particular tradition. We do this most of the time unconsciously, especially when we are very little and learn through imitation. There is some controversy here concerning this ability to mimic. Can non-human animals imitate each other? Some authors think that they can, especially primates. But others think that this is unlikely. Experiments have produced contradictory results. In any case, to imitate someone else, you need to be able to figure out his or her intentions. I can imitate another human being, I can even imitate an animal (to which I have attributed an intention), but I cannot imitate non-intentional beings. It doesn't make sense to 'imitate' a machine, for instance, unless I personify it (a mime, for instance). Think of little children 'imitating' a train.

Because traditions are cultural knowledge, the intentions we grasp when we learn a tradition are not only the intentions of the person who is teaching us, but of all those who have taught that tradition since the very beginning. We can compare this to trying to understand how a car works, for example. I am not only trying to grasp the intentions of the car designer, but of all those who have contributed to making that car, whose intentions have been 'congealed' in the actual pieces of the machine. To the extent that we accumulate within our minds all the cultural knowledge that we have learnt since we were born, we can safely say that 'understanding a human being' is understanding a world.

A parallel can be drawn with genes and genetic knowledge. Except for identical twins, we are all genetically unique. But in fact, the number of DNA sequences not shared by two individuals who are not close relatives is very small, between 0.1 and 0.2 percent of the complete genome. All the rest is shared with other human individuals, and about 98 percent is also shared with non-human animals, in different proportions, according to how far from us they happen to be in the evolutionary scale. When we look at the genes of an individual organism, we are in fact looking at the genes of a whole gamut of different individuals. Exactly the same thing happens with cultural knowledge. By looking at the cultural knowledge of a particular individual, we are actually looking into the minds many different individuals.

Notice now that to understand anyone's intentions, we need to see ourselves in them. This is what ToM is all about actually. When I can figure out someone else's intentions it means that I can somehow 'represent' her mind in my mind; in other words, I can see myself in her. We do this all the time when we are communicating with other people and trying to understand what they say. You are doing it right now while reading this book. Here is how Austrian philosopher Ludwig Wittgenstein puts it:

> We also say of some people that they are transparent to us. It is, however, important as regards this observation that one human being can be a complete enigma to another. We learn this when we come into a strange country with entirely strange traditions; and, what is more, even given a mastery of the country's language. We don't *understand* the people. (And not because of not knowing what they are saying to themselves.) We cannot find ourselves in them (*Wir können uns nicht in sie finden*).

Wittgenstein is describing perhaps the most troublesome consequence of cultural diversity: cultural misunderstanding. Why is it that 'we cannot find ourselves in them'? Because their minds are an enigma to us. This is so not only because we don't know the particular experiences they have gone through as individuals, we do not know their individual knowledge, but also because the myriad individual intentions objectified in their cultural behaviour are also unknown to us. We do not understand them as individuals because we do not understand their world and, therefore, we cannot find ourselves in them.

So this is what 'understanding a culture' is all about. Later on we will see how we anthropologists go about trying to do that. But now let us go back to the problem of behaviour. How does 'understanding another individual' help us to anticipate his or her behaviour? Once again, the comparison with genetic and individual knowledge is instructive. As soon as we know the genetic knowledge of a particular organism, we can predict with fair accuracy how this organism will react to a certain stimulus. We have seen that predictions are never 100 per cent sure because genes are not 100 per cent effective. Predictions based on individual knowledge are definitely more complex because we need to know the experiences the animal has gone through and how its organism reacts to those experiences. In normal circumstances, it means that an animal's non-instinctive behaviour is practically impossible to predict. But we can work around that difficulty by training the animal, as I have already pointed out. In this case, it is a particular set of experience which eclipses, as it were, all the others so predictions based upon individual knowledge can become almost as accurate as those based upon genetic knowledge. When we have trained an animal to react in a particular way to a particular stimulus, we can predict its behaviour under these particular circumstances with a fair degree of accuracy.

Can we produce the same sort of predictions for human cultural behaviour? At first sight, unpredictability seems to be the norm here. But if you take a second look, you will see that this is not exactly true. For instance, I can predict that you will not jump off of your chair at this very moment and slap the first person you see. You haven't done it. Have you? Can I predict human behaviour or not? You may point out that mine was a negative prediction; it is much easier to predict what others will not do than what they will actually do. Still, our day-to-day life would be a nightmare if we could not predict other humans' behaviour with a certain degree of accuracy. When you attended a conference, say, on international politics last week, you predicted that the speaker was going to give you a lecture on international politics. Even though you didn't know exactly what he or she was going to talk about, you could be pretty sure that the lecture would not be on quantum physics, for instance. How did you know?

Partially, these predictions are based on individual knowledge. This has never happened before, so why should it happen now? Statistically, something that has never happened is unlikely to happen and something that repeatedly happens is likely to happen again. Lecturers at the university always come to the classroom and give a boring lecture; they have done this before many times, so they are very likely to do it again. You might be wrong, you might come across a lecturer who gives an interesting lecture, but it is unlikely. Drivers always drive – in this country at least – on the right hand side of the road. You might come across someone who drives on the left hand side of the road, but it is, for mere statistical reasons, unlikely. But there is more to it than that.

The first time you attended a lecture at your university, you didn't have any previous experience of what lectures were like, and probably lots of things happened to you that you didn't anticipate: you didn't foresee that the lecture would be boring because you didn't have prior individual knowledge of lectures. But your cultural knowledge about universities and lectures would have allowed you to predict many things about the lecture without having had any previous experience. You could have predicted that a lecturer would give you a lecture on ancient history, for instance, and would not jump on top of the table and start to dance, because that would not make any sense according to the cultural knowledge that you already had in your mind. Someone who might not have been brought up in that particular culture would understand a lecturer jumping on top of the table and starting to dance in a totally different way.

Making sense of other people's behaviour

Take note of this very important concept of 'making sense'. We do things because they make sense; we don't do other things because they don't make any sense. Cultural behaviour is behaviour that makes sense among the people who share that cultural knowledge. 'Making sense' is another way of talking about 'finding yourself in the other'. I can make sense of another person's behaviour because, as it were, 'I can see myself behaving in that particular way'. Remember, once again, that what we call cultural knowledge, as embodied in ideas, traditions, objects, texts, etc. is no more than a set of congealed intentions. In the same way as I can't understand another person if I can't figure out his or her intentions, I can't understand another person's culture if I can't figure out that set of congealed intentions.

A behaviour that makes sense is not, in any conceivable way, an obligatory behaviour. And that is perhaps the most remarkable difference between behaviour as determined by cultural knowledge, on the one hand, and that determined by genetic and individual knowledge, on the other. Behaviour as determined by our genetic knowledge is relatively compulsory. Like all animals, we feel impelled to behave according to our instinctual drives. Behaviour as determined by our individual knowledge is equally compulsory, but it is much more uncertain than instinctual behaviour, in so far as it is hard to ascertain the set of experiences that compose the individual knowledge of a particular subject, except when that subject has been trained to perform a particular task. In this latter case, animals feel equally impelled to behave according to their training (individual knowledge) and to instinctual drives (genetic knowledge). The compulsion that our cultural knowledge exercises on us is much more subtle. We also feel compelled to behave according to our cultural knowledge, which means that we feel compelled to behave in a sensible way (unless we suffer from a psychological disorder). But notice how peculiar this compulsion is: we may follow it or not. We may obey a legal prescription or we may not. If we were to obey all legal prescriptions in the same way as a trained dog obeys its master's orders, we would not call that behaviour 'obeying a legal prescription'. This degree of choice explains why cultural behaviour appears to be totally unpredictable at first sight. But in reality, we could not survive a single day if we couldn't predict other people's behaviour. The fact that we can predict others' behaviour indicates that culture enacts a special form of compulsion upon us, the *compulsion of meaning*.

In other words, culture does not obligate us to behave in a particular way; rather, it obligates us to behave in a way that makes sense to the people who share that culture. In so far as the set of behaviours that make sense in a particular society is limited, I can predict cultural behaviour. I can predict that behaviour that falls outside that particular set is rather unlikely among the people who live in that society. But, in so far as a particular set of culturally accepted behaviours includes an infinite number of behaviours, cultural behaviour turns out to be unpredictable. So here is another paradox of culture: in all cultural behaviour there is always a trade-off between predictability and unpredictability.

Imagine the set of cultural behaviours available to any one person in a particular society as the infinite set of prime numbers. If I know that a number is a prime number and nothing more, there is no way for me to know which prime number it is, because the set of prime numbers is infinite. Yet, the set of prime numbers is a 'smaller' infinite than the set of natural numbers, for instance. Similarly, the set of all culturally acceptable behaviours in a particular community is infinite, but it is a smaller infinite than the set of all the possible behaviours, both culturally acceptable and unacceptable. This makes it impossible for me to fully predict cultural behaviour.

In fact, in any human interaction there is always a complex relationship between the predictable and the unpredictable. A totally predictable conversation would be utterly useless. If you could predict exactly all the words that I am going to write, reading this text would be a waste of time. Conversely, talking to someone who makes totally unpredictable statements is also useless because you can't make any sense of what he or she is saying.

In this chapter we have been looking at the most distinctive characteristic of the human species: cultural knowledge. We have seen how it emerged and what its main characteristics are, specifically in comparison with the other forms of knowledge we humans share with non-human animals. In the next chapter we shall be looking at the ways in which anthropologists go about trying to explain culture. But what does 'explaining a culture' actually mean? Why should we explain a culture? We already know it: under certain circumstances, a culture can become an enigma to us.

Bibliographical note on Chapter 2

A traditional anthropological view of culture can be found in Geertz's (1973) classic work, which is particularly relevant in terms of the problem of meaning and interpretation. More modern accounts of the concept of culture, still within the more or less traditional anthropological perspective, are Carrithers (1992), Kuper (1994) and Mahler (2013). The first two are more academically oriented while the third is addressed to a wider audience. Kronenfeld (2018) recent publication, which is also academically oriented, provides an updated overview of the culture concept according to the American tradition of cognitive anthropology. An early evolutionary perspective on culture that tries to come to terms in a systematic way with the relationships between culture and human biology can be found in Boyd and Richerson (1985). This is a theoretically sound work with far-reaching effects in later research on cultural evolution, as we shall see in Chapter 4. It is not an easy read, however, for those without some mathematical training. As for the evolutionary origins of culture, Tomasello's work (1999) constitutes an excellent introduction and Sterelny (2012) provides an updated synthesis specifically focused on the origins of cultural learning. The differences between cultural learning and social learning are explored in Tomasello et al. (1993). Harris (2012) deals with this very same problem of cultural learning but from a developmental perspective. See also Morin (2016) for a modern, interdisciplinary and very original approach to cultural reproduction. But perhaps the most comprehensive account of the crucial role that culture has played in human evolution is without a doubt Henrich (2015), a highly readable text. Written from a modern evolutionary perspective, the work includes a vast array of historical, archaeological, primatological, paleoanthropological, psychological and ethnographic information. The concept of emergence, which I have used to refer to both to mind and culture, originated in philosophical discussions about the concept of consciousness and the so-called mind-body problem (see Clayton and Davies 2006). An application of this concept to the analysis of mind and culture can be found in Deacon (2012), a work full of deep insights but with a particularly dense argument that makes it inappropriate for beginners. Important works focused on the origins of language and the specificity of human communication are Deacon (1998), Scott-Phillips (2015) and Tomasello (2010). Deacon pays special attention to the symbolic nature of human communication, Scott-Phillips emphasises what he terms its 'ostensive-inferential' character (that is, the fact that humans communicate intentions rather than information) and Tomasello underscores the inherently cooperative nature of all human communication. Another classic text that tries to build bridges between traditional anthropological approaches to cultural knowledge and more modern cognitive and evolutionary theories is Sperber (1996) and, from a different theoretical perspective, Strauss and Quinn (1997).

References

Boyd, P. and P.J. Richerson. 1985. *Culture and the Evolutionary Process*. Chicago: Chicago University Press.
Carrithers, M. 1992. *Why Humans Have Cultures*. Oxford: Oxford University Press.
Clayton, P. and P. Davies. 2006. *The Re-emergence of Emergence*. Oxford: Oxford University Press.
Deacon, T. 1998. *The Symbolic Species*. New York: W.W. Norton.
———. 2012. *Incomplete Nature*. New York: W.W. Norton.
Geertz, C. 1973. *The Interpretation of Cultures*. New York: Basic Books.
Gould, S.J. 1991. 'Exaptation: A Crucial Tool for Evolutionary Psychology'. *Journal of Social Studies* 47: 43–65.
——— and R.C. Lewontin. 1979. 'The Spandrels of San Marco and the Panglossian Paradigm: A Critique of the Adaptationist Programme'. *Proceedings of the Royal Society of London* 205(1161): 581–598.
Harris, P.L. 2012. *Trusting What You Are Told*. Cambridge, MA: Harvard University Press.
Henrich, J. 2015. *The Secret of Our Success*. Princeton, NJ: Princeton University Press.
Kronenfeld, D.B. 2018. *Culture as a System*. London: Routledge.
Kuper, A. 1994. *The Chosen Primate*. Cambridge, MA: Harvard University Press.
Mahler, S.J. 2013. *Culture as Comfort*. Boston, MA: Pearson.
Morin, O. 2016. *How Traditions Live and Die*. Oxford: Oxford University Press.
Scott-Phillips, T. 2015. *Speaking Our Minds*. New York: Palgrave Macmillan.
Sperber, D. 1996. *Explaining Culture*. Oxford: Blackwell Publishing.
Sterelny, K. 2012. *The Evolved Apprentice*. Cambridge, MA: MIT Press.
Strauss, C. and N. Quinn. 1997. *A Cognitive Theory of Cultural Meaning*. Cambridge: Cambridge University Press.
Tomasello, M. 1999. *The Cultural Origins of Human Cognition*. Cambridge, MA: Harvard University Press.
———. 2010. *Origins of Human Communication*. Cambridge, MA: MIT Press.
———, A.C. Kruger and H.H. Ratner. 1993. 'Cultural Learning'. *Behavioral and Brain Sciences* 16(3): 495–511.

3
THEORIES OF DIFFERENCE

Understanding human diversity

Human diversity originates in our cultural knowledge. Because cultural knowledge can only be shared by a group of people, and cultural knowledge can accumulate increasingly complex cultural objects almost indefinitely, the cultural objects of different culture-sharing groups can become mutually unintelligible. Cultural diversity generates misunderstanding. Notice that cultural diversity differs from linguistic diversity. As a result, linguistic misunderstanding also differs from cultural misunderstanding. For example, I visit an English-speaking North American indigenous community. They speak English to me so I understand their language, but I don't know what they are talking about most of the time, because their culture is alien to me. If I visit Greece, I can't communicate with the people or understand what they say, because I don't speak Greek. But their ways of life are familiar to me. In this instance, a lack of linguistic understanding does not imply cultural misunderstanding. A translator would help me understand the Greek of my hosts. But what sort of 'translator' do I need when I don't understand the culture?

We already know how cultural misunderstandings originate. We will now see how anthropologists 'solve' these misunderstandings, by trying to translate one culture in terms of another. Learning a foreign culture the way anthropologists do is different from learning one's own culture, even though there are undoubtedly many parallels. First of all, when learning a foreign culture an anthropologist does not want to become a 'native' of that culture (which isn't possible anyway). He or she is simply trying to understand the locals, rather than trying to 'live' their lives. Think of what understanding another person means. (Does it mean to 'live' his or her life?) Let us have a quick look at the methods and research techniques that anthropologists use to learn a foreign culture. Ethnography is the method anthropologists use to learn a different culture and translate it into the terms of another.

In fact, this translation process occurs simultaneously to the learning process, since you can only understand a particular foreign culture once you have been able to translate it into yours. Participant observation is the technique used to gather ethnographic data.

Ethnography and participant observation can be defined as an attempt to emulate the way that children learn their own cultures. Over a relatively long period of time, anthropologists live with the people whose culture they are trying to learn, observing and participating in their day-to-day interactions, and sharing their experiences. In the end, we can 'see ourselves in them', but we do not ourselves become 'native'.

We have defined culture as a system of meanings and symbols in terms of which humans govern their behaviour. Anthropologists can find these systems of meanings in two places: in people's minds and in their behaviour. We know that cultural knowledge originates in an individual's knowledge of another individual's knowledge of another individual's knowledge, etc. We also know that all forms of individual knowledge are stored in the minds of those individuals who possess that knowledge. The same logic seems to apply to cultural knowledge. In fact, late nineteenth and early twentieth century anthropologists used this principle to study the cultures of so-called 'primitive' peoples. They visited villages and asked elders to report everything they knew about their cultures (with the help of an interpreter when necessary). This approach depended on the conviction that when a person grows up in a particular culture, his or her mind becomes a repository for cultural information. So if you want to learn about the culture of particular society, just go there and ask someone, preferably old people, since the older they are the more culture will have been stored into their minds.

Certainly, the most elementary way of finding out the contents of someone else's mind is to ask him or her. But we can only learn so much by asking, because much of our cultural knowledge is unconscious. We have already seen that much cultural knowledge is embodied in our habits, in our ways of doing and not doing things, and we might not even be aware that we have that knowledge. How can the locals (what anthropologists call 'informants') tell something if they aren't even aware that they know it? What would anyone say if an anthropologist from another society asked him or her, 'How do you relate to your father, your mother, your siblings, your children and your in-laws?'. The person probably would not know how to respond, unless he or she happened to be going through some kind of conflict that caused him or her to think about these relationships. Ordinarily, people do not think about how they relate to their close kin; they just do it. Furthermore, conscious understanding of a particular relationship might differ significantly from the way that relationship actually is. Let us return to the example of gender. Our gendered behaviour is usually so deeply embedded in our nervous system that we find it absolutely natural. At the same time, in Western societies we have all learned that men and women are equal, and that, for example, husbands and wives should share domestic chores equally. When anthropologists ask couples from Western societies how they organise domestic chores, the couples normally respond that they share tasks fifty-fifty. Yet observation shows that these very same women do most of the housework themselves, while their husbands watch TV. How can we explain this phenomenon? Another limit concerning what people explicitly tell an anthropologist about their culture relates to how much they trust the anthropologist. Again, imagine a foreign anthropologist who is a total stranger asking a native about her family. The native would probably tell the anthropologist very little, not because she were hiding things, but because she would feel uncomfortable sharing that information with a stranger.

We only need to watch what is going on in a particular society to make up for all the deficiencies of the previous method. If culture is a system of meanings and symbols in terms of which people govern their behaviour, we can infer the meanings and symbols by observing their behaviour. Needless to say, we learn most when we ask questions at the same time. Dialogue and observation complement each other rather than being mutually exclusive. This leaves the question of how anthropologists can observe the behaviours that interest them. Many behaviours can be observed in public places; we only need to look around and see what people are doing. But how do we observe behaviour in private places? Anthropologists gain access to such places by interacting with people and establishing close relationships with them. Ethnographic research is not qualitatively different from the way we obtain the knowledge we have of other people in our day-to-day lives. How do we manage to understand the thoughts and behaviour of other people? By interacting with them, *living with* them. In ethnography, daily life becomes not only the object but also the main means of research. The only difference is that in ethnography we interact with people deliberately and systematically.

A comparison between ethnographic research and psychological research can be instructive. Both anthropologists and psychologists study human behaviour, but very rarely will psychologists conduct ethnographic research. Instead, psychologists usually obtain their information on human behaviour from two sources. First, by asking them to perform psychological tasks in a laboratory. These people are normally psychology or other university students, and they normally receive a small payment. Alternatively researchers place ads for participants fitting a particular description (e.g. people older than a certain age, or from a particular ethnic background). Psychotherapy provides a second source of information. In this case, psychologists do not need to recruit people to visit the laboratory. Rather the people themselves come to the psychologist asking for help. Like a medical doctor, the psychologist obtains information by talking to patients and, occasionally, administering psychological tests. In this case, rather than receiving a payment from the psychologist, the patient pays the psychologist for his or her help.

These differences between ethnographic and psychological methods have to do with the different objectives of anthropology and psychology. Psychologists are not particularly interested in human diversity; rather what interests them is the human mind, or perhaps the human 'psyche', if we understand by that the universal characteristics that all human minds share. We all have been brought up in different contexts, but this does not make our minds (our 'psyches') substantially different.

It makes them look different, certainly, but psychologists hold that these are superficial differences that mask an underlying common human nature, a 'truer' essence that reveals itself once we put aside those apparent cultural differences. Think of a medical doctor trying to find out about a patient's illness. First the doctor listens to how the patient describes his or her sufferings, the symptoms. Depending on the patient's education, background and culture, this description can be very different. The doctor must put aside all these differences and try to uncover the underlying disease, the real cause of the patient's sufferings behind his or her precise or imprecise description of the symptoms. The patient's cultural background might make a difference in the way he or she describes (and experiences) the disease, but not in the disease itself, which is what the medical doctor wants to find out. Much the same happens when psychologists try to understand what is going on in people's minds. While the medical doctor can use biomedical analyses and observations (blood tests, X-rays, etc.) to assess the condition of his or her patient's health, which will provide him or her with information irrespective of the way the patient himself or herself describes his or her own symptoms, the psychologist uses psychological tests to obtain similarly objective information about the individual's mind or mental health (in the case of psychotherapy). The underlying assumption is that there is a common universal human nature that enables medical doctors to find the ultimate causes of human illnesses and for psychologists to find the ultimate causes of human behaviour.

Anthropologists' perspective is rather different. For a science that is interested in understanding human differences, a methodology that consists in precisely setting aside those differences is not of much use. How can those differences be revealed? Cultural behaviour is always behaviour *in a context*, for cultural differences are always differences in the contexts in which people have been brought up. Of course that cultural knowledge needs to be somehow 'uploaded' into our minds if it is to have any effect upon human behaviour. But the (potential) advantage of cultural knowledge in evolutionary terms is that it makes people behave in a certain way under certain conditions. Natural selection can only act upon behaviours, not upon ideas or mental representations. Notably, some people can be self-reflective concerning their own cultural knowledge. But most of the time we are not – again, think of the unconscious cultural knowledge we saw before. So we cannot 'download' that cultural knowledge at will, as if we were downloading information from a website. Remember the anthropologist trying to find out about our family relationships; the best way to learn these answers is to observe people while they are interacting with their family members. One reason for this fact is that cultural knowledge is always collective knowledge.

I get my cultural knowledge from someone else, who in turn got it from someone else. The source of cultural knowledge is always a group of people living under specific conditions, with loosely defined space-time boundaries. We must turn to this group of people living under these conditions if we want to understand their culture. The anthropologist does not have to interact with every group member, an impossible feat considering that cultures are not tightly circumscribed social units.

Occasionally, we might find cultural groups with clear-cut limits, such as (relatively) isolated tribal societies or religious sects. But most of the time cultures endlessly overlap each other. And even if we could identify a delimited cultural group, as we saw earlier, we would not find two individuals within that group with exactly the same cultural knowledge; for cultural knowledge originates in social interaction and no two individuals will have exactly the same interactions throughout their lives in any group. Again, this is very similar to what we found with genetic knowledge: we all share genes but nobody except identical twins has exactly the same genes.

Still, someone might argue that in order to analyse the cultural knowledge of a particular individual – even if that cultural knowledge is collective knowledge – clever anthropologists could have devised a test that would enable them to download that cultural knowledge in a lab, in much the same way as psychologists obtain their information from the individuals they study. True, there is a branch of anthropology called 'experimental anthropology' that does precisely that and that has yielded some interesting results. But there is a problem with this type of research as applied to cultural analysis. If we bring people into a lab and make them go through psychological tasks, their cultural knowledge is likely to be severely distorted. No one has been brought up to go into a lab and take psychological tests, so people's reactions under those conditions, whatever else they happen to be, cannot be seen as genuinely 'cultural' reactions. That explains why we anthropologists need to do ethnographic research, that is, observe people in their natural environments, in their natural settings; experimental research needs to remain secondary to ethnography.

Ethnography is the documentation of cultural differences. But once we have documented those differences, how do we explain them? Stated otherwise, once we know why humans have cultures, we need to know why particular societies have this particular culture and not that other one. Anthropologists have devised all sorts of theories to account for human diversity. But theory and method in anthropology are closely related. As we shall see in what follows, the constitution of ethnographic fieldwork and participant observation as distinctive features of anthropological research goes hand in hand with the development of leading theoretical approaches in the history of anthropology. Remember that we anthropologists are cultural interpreters, so explaining cultural differences is putting one culture in terms of another.

Opening approaches: anthropology as history

We shall be looking at four very traditional and well-known schools of anthropological thought: social evolutionism, historicism, functionalism and structuralism. Certainly, in the history of anthropological thought there are more schools than these four. But I have chosen them because they can be taken as representative of particular patterns of thought, and shared with many more anthropological theories, in the analysis and explanation of cultural differences. To synthesise the history of anthropology in just four theoretical approaches is, needless to say, an over-simplification. But, hopefully, it will enable us to get at the core of what explaining culture is all about.

Social evolutionism

Anthropology became an academic discipline in the second half of the nineteenth century. At that time, Western societies were haunted by the idea of progress, of getting rid of the shackles of tradition and advancing toward the future, which was thought to be very different from the past. Many of those societies had just gone through industrial and democratic revolutions. They were changing very rapidly and those changes were seen as good and necessary. The intellectual legitimation of this somewhat volatile situation came from a social theory known as social evolutionism. As we shall see, this brand of evolutionist thinking was very different from the biological evolutionism espoused by Darwin around the same time. It originated not in Darwin's ideas but in the social thought of the Enlightenment philosophers of the previous century. Social evolutionists held that human societies changed through time for the better. And those changes were the result of the progressive accumulation of knowledge that we know as 'culture'. A clear parallel was drawn between the development of the individual and the development of society. Individuals are born with no knowledge but as they grow up, knowledge pours into their minds, so that when they are old they have plenty of it. The same applies to human societies. Culture is the knowledge of human societies, and history is the name we give to their development process. In the same way as any individual acquires more knowledge as he or she grows older, human societies acquire more culture as they advance through history. This is what we call 'progress'. Otherwise stated, in the same way as a child has less knowledge than an adult, early societies (what at that time were called 'primitive societies') had less culture than modern societies.

This way of looking at culture is very intuitive. If culture is social knowledge, then as societies go through history they accumulate culture, in the same way as an individual accumulates knowledge as he or she goes through life. But how could this theory help us explain human cultural diversity? Social evolutionists did not think that there was such a thing as cultural diversity or plurality as we understand it now. They always used the word culture in singular: for them, there was only one single culture for the whole humankind. Consequently, human diversity does not originate in the fact that humans have different cultures, but that some have more culture than others. Thus, tribal and 'savage' societies (which from then onwards started to be known as 'primitive') do not have a different culture from ours; they simply have less culture than us because they find themselves in a previous state of historical development. Again, the parallel with individual development is crystal clear: you don't say that a small child who does not know how to perform long division has a different 'culture' from the adult who knows how. Rather, you understand that the child has not acquired the knowledge one needs to be able to perform division.

It was not only tribal or small-scale societies that were seen as primitive, as lacking culture or having less culture than us, but any society that substantially differed from Western societies would also be seen as less advanced or underdeveloped. And even within Western societies themselves, populations could also be sorted out into more or less advanced, that is, with more or less culture, according to how different they were from the ruling elites. That is how social evolutionists explained human diversity: to put it crudely, anyone who is different from me is inferior to me.

Figures 3.1 and 3.2 show in a very simple way the underlying assumptions of social evolutionism. Human societies accumulate culture as they go through history. History is not a chaotic and contingent concatenation of events but a necessary process determined by the laws of cultural evolution, namely, the progressive accumulation of knowledge through time. So history must be the same for all humankind because culture, the product of history, is the same for all humankind. But not all human societies find themselves at the same stage of historical development, nor do all societies go through history at the same speed. Therefore, not all of them have accumulated the same amount of cultural knowledge. The fastest ones are the most advanced; the slowest are the most primitive. By this logic, human societies could be sorted into more advanced and less advanced, according to the amount of history they have gone through, which in turn determines the amount of culture

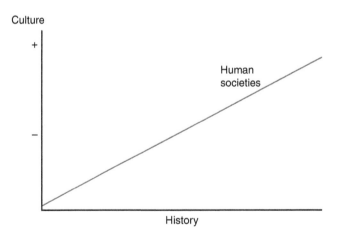

FIGURE 3.1 Evolution of human societies

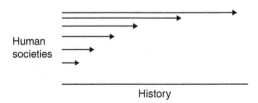

FIGURE 3.2 The race of history

they have accumulated. At any one point in time we could deploy all human societies alongside the diagonal in Figure 3.1. Those with the most history and culture (the 'advanced') would be in the top right corner, while those with the least (the 'primitive') would cluster in the bottom left corner.

For social evolutionists, going through the spectrum of human diversity was like time travel. Any society that happened to be different from Western societies was thought to occupy a previous stage of historical development. Thus anthropology began as a historical science of sorts, a science whose purpose was to reconstruct a particular period of history. The difference between anthropologists and ordinary historians was merely methodological. It had to do with the particular way anthropologists reconstructed history: instead of using written records or archaeological remains, they would view current 'primitive' societies as living fossils. This method turned out to be particularly handy for the reconstruction of very remote periods of human history, so remote that no records, written or otherwise, had survived. But in fact it could be used for the reconstruction of any historical period, as long as there was a society in the present that found itself at that particular stage of development.

Many have seen a connection between this branch of evolutionist thinking and contemporary Darwinian biological evolutionism, and some even go as far as postulating that social evolutionism emerged from Darwin's influence on the social sciences. Nevertheless, the differences between social and biological evolutionism are very noticeable. Both kinds of evolutionism aim at explaining a form of diversity: biodiversity in the case of biological evolutionism, cultural diversity within the human species in the case of social evolutionism. But commonalities between the two practically stop there. As we have seen in the previous chapters, Darwinian evolutionism is based upon the key concepts of natural selection and adaptation. Current biological species are diverse as a result of the different adaptive challenges faced by their ancestors. But none of them could be seen in any way as being 'more evolved' than the others. All species are, generally speaking, more evolved than their respective ancestors but not more than any of their contemporaries; that would not make any sense in Darwinian thought because by definition all species are adapted to their own EEA. But the notion that some peoples are more evolved than their contemporaries is precisely the central tenet of social evolutionism. Natural selection and adaptation are totally alien concepts to nineteenth century social evolutionism. Some later offshoots of social evolutionist thinking in the twentieth century applied a logic of biological evolutionism. But their nineteenth century forebears mostly overlooked or simply ignored Darwin's thought. For these initiators of evolutionist thinking in the social sciences, diversity within the human species originates in the different speeds at which each society evolves (i.e. accumulates cultural knowledge through history), not in the need to adapt to different environments. For biological evolutionism, at any one point in time, no species can be said to have evolved more than the others; for social evolutionism, by contrast, that is precisely what evolution actually means as far as human societies are concerned.

Now anthropologists like to think that modern social and cultural anthropology, as it developed in the twentieth century, began as a critique of social evolutionism. In reality, the three theoretical approaches that we will be looking at next, historicism, functionalism and structuralism, took the critique of nineteenth century social evolutionism as one of their foundational theoretical arguments. However, it is worth remarking on how pervasive social evolutionist thinking still is, especially outside academic circles and even among some social scientists. The idea that human societies can be classified into 'developed' and 'underdeveloped', and that the only problem with the latter is that they lack what the others have already achieved is by no means strange or unusual. By this logic, the future of the so-called underdeveloped world is, almost by definition, to become developed; this idea appears frequently in present-day understandings of human diversity.

In a way, despite modern anthropologists' misgivings, it could be argued that some form of social evolutionism is not too far-fetched. Undoubtedly, cultural knowledge is accumulated through history, at different speeds in different societies. This does not mean though that Western societies must necessarily be at the pinnacle of the cultural evolutionary process, or that the only way out of their alleged underdevelopment is for non-Western societies to follow the path blazed by the West. The comparison with individual development, or rather with individual acquisition of knowledge (cultural or individual), may turn out to be instructive once again. Even if we dismiss right from the start the analogy between non-Western societies and children (because they are supposed to have 'less culture' than us), there is a way in which the distribution of knowledge among a group of individuals is perfectly commensurate. In other words, even though each individual human being might proceed along a path of cognitive development in his or her own particular way, at his or her own speed, etc. and none of these particularities makes him or her more or less clever than the other individuals, all these individuals are capable of interacting and exchanging the knowledge they have. Thus, at any point in time we will always have individuals who have more knowledge about any particular subject and others who have less. So perhaps human societies could be seen along similar lines. To some extent, each society follows its own path of cultural evolution and no one can be judged as being more or less evolved than any other. But at the same time, since many forms of cultural knowledge are fully commensurate, not all societies may possess the same amount or the same quality of that cultural knowledge.

Be that as it may, social evolutionism was the first theory in the history of the social sciences that tried to account for human diversity in cultural terms. Social evolutionists held that all humans were biologically and psychologically equal, by which they meant that differences resulted from enculturation rather than from innate qualities. This idea is sometimes referred to as the principle of the psychic unity of humankind. According to this view, we are all born equal, and what makes us different is the cultural knowledge that we acquire as we grow up. If we have been lucky enough to be born in an 'advanced' society, we will get plenty cultural knowledge, and of the right kind. If we have not been so lucky and we happen to be born in a 'primitive' society or a 'backward' social group, we will not get much cultural knowledge, and the little we get may not even be right.

Historicism

Modern anthropologists see social evolutionism as a first attempt at providing a systematic account of human diversity. But evolutionists failed to take into account a crucial component of cultural difference: the actual existence of *different* cultures. Otherwise stated, the word 'culture' should not be used in the singular but in the plural. Cultural differences cannot be explained as simply the result of having more or less cultural knowledge, for they have their origins not in quantitative but in qualitative variations in the very nature of that cultural knowledge. The first critique of evolutionist thinking in anthropology came from the United States, from a school of anthropological thought known as 'historical particularism', initiated by one of the founding fathers of modern anthropology: Franz Boas (1858–1942), a German anthropologist who made his career in the United States.

Interestingly, two of the main theoretical challenges to evolutionist thinking in anthropology, historicism and functionalism, began as methodological revolutions rather than as properly theoretical confrontations. It was in the actual process of collecting data on human cultural behaviour that the weaknesses of social evolutionism first came to the fore. Boas is not only considered the founder of modern American cultural anthropology but he was also one of the first anthropologists to study Native American societies through systematic direct contact with their members. Before him, anthropologists, that is, mostly evolutionist anthropologists, were not terribly concerned with interacting with the members of the societies they wanted to study. Though some of them did actually visit those societies, the majority relied on second hand information provided by travellers or anyone who happened to be living near them. These were known as anthropologists' 'correspondents'. It is worth emphasising the contrast between these two different ways of gathering information in anthropological research, for much of the theoretical divergence between social evolutionism and modern anthropological theory builds upon that fundamental methodological contrast.

We have already seen the importance of the ethnographic method for modern anthropological research. But nineteenth century evolutionist anthropologists, the so-called 'pre-modern' anthropologists, did not gather information in this way. Remember that evolutionist anthropologists were mostly history theoreticians: they approached cultural diversity by producing a theory of social evolution that could explain at what stage of historical development particular societies found themselves. To this effect, they certainly needed data on the cultural knowledge characteristic of different societies. But the final interpretation of that information could only be made by comparing the cultural knowledge of different societies in order to sort them from least evolved to most evolved. This is known as the 'comparative method'.

Now it is clear from this that anthropological knowledge as such, that is, the explanation of cultural differences, originates not so much in the analysis of particular societies with their particular cultures (or more accurately the particular bit of universal cultural knowledge that they have managed to produce, for don't forget that the concept of 'cultures' in the plural was totally alien to evolutionist thinkers), but rather in the comparison of different societies, the more the better, because only by collecting lots of cultural information from lots of very different societies could a sound theory of historical development be set forth. Consequently, a distinction should be drawn in evolutionist anthropological research between the theoretician (in charge of analysis and theory) and the field worker (or anthropologist's correspondent, in charge of data collection). These latter did not have to be professional anthropologists (most of them were not). The only condition they had to meet was to live or to be in close contact with a particular society of interest to anthropologists. So the majority of correspondents were missionaries, traders, travellers, functionaries of the British Empire or Indian agents in the United States. The blending of the roles of theoretician and field worker marked the end of pre-modern anthropology and the beginning of what we now understand as modern anthropological research.

Let us get back to Franz Boas, the main character of our story in this section. Boas was not an anthropologist by training but a physicist and a geographer, and he initially went to Baffin Island in the American Arctic to study, among other things, inhabitants' colour perception. During this fieldwork, he made friends with the locals and became interested in their culture and ways of life, so he decided to become an anthropologist. Let me emphasise how important the direct contact with local people was, right from the beginning, in accounting for Boas's interest in anthropology. This is clearly a very different initiation into the study of cultural difference from that of the majority of evolutionist anthropologists. It was from this bottom-up perspective that Boas realised that theories of social evolutionism were not appropriate for making sense of the sort of information he had been gathering. This is to be expected, given that Boas did not collect a small amount of specific information from numerous societies, but rather a vast amount of in-depth information from one single society (or a handful of societies at most). This is the sort of information that can be gathered by one single individual rather than multiple correspondents. As a result of that, Boas became very sceptical of the grand theories of historical development put forth by evolutionist anthropologists. Those theories were of no help to him given the kind of research he had been conducting. First of all, social evolutionists defined primitive peoples as ignorant: what made them different from us is the fact that they had less culture than us. But Boas did not see that the people he was interacting with were ignorant at all in any way. How could we account for that difference? Was it the environment perhaps? Cultures may be different because of the different environmental challenges they had to face, as perhaps a modern adaptationist would be tempted to argue. But Boas discarded environmental determinism since he did not see a one-to-one relationship between the characteristics of the environment and the characteristics of particular cultures.

He proposed instead a new theoretical approach that would focus, not on the comparison of the cultural achievements of different societies, but on understanding particular people in their particular historical and cultural milieu.

As evolutionist anthropologists had argued, Boas also saw culture as a historical product, as the product of the accumulation of knowledge through history. But Boas's conception of history was very different from that of social evolutionists. The explanation for why a particular people have the cultural beliefs and practices that they have has nothing to do with alleged universal laws of historical development, as social evolutionists had postulated. For Boas and the Boasians, each particular society has its own cultural traditions. And the only possible explanation for those cultural traditions is historical. This historical explanation refers not to some general historical trajectory common to all humankind, as evolutionist anthropologists would have it, but to the particular history of each particular people. It refers not to some necessary laws of historical development, such as the progressive accumulation of cultural knowledge, but to the contingent events that comprise the particular histories of particular peoples. In other words, there is no such a thing as a general history of humankind divided up in different stages, which in turn correspond to different stages of cultural evolution. Rather, there are only particular histories of particular societies. Once history is no longer seen as a necessary process, there cannot be one single history for the whole of humankind. Rather each society has its own history. And since culture is the product of history, multiple histories give rise to multiple cultures. There is not a unique culture, in the singular, common to all humankind but myriad particular cultures, in the plural, each one resulting from the particular history of each particular people. In Boas's understanding of cultural knowledge, cultures end up being incommensurable: you cannot think of one culture as being 'more' or 'less' than any other. It is hard to think of a more radical reversal of evolutionist principles.

What could be the aim of anthropological research within this new theoretical framework? Certainly, not the formulation of general theories of historical development and cultural evolution. Culture is the object of knowledge of anthropology, but culture can be seen both as *explanandum* and *explanans* in Boas's theoretical project. In other words, culture is both something that needs to be explained which, in turn, explains something else, namely, human behaviour. From the first point of view, culture is a historical product and, as we have seen, it can only be explained by reference to the particular history of the particular people who have that culture. So explaining culture entails a process of historical reconstruction. Again, a superficial similarity with social evolutionism seems to insinuate itself. Evolutionists also held that anthropologists were like historians, whose purpose was also the intellectual reconstruction of history. But notice the difference between Boas' task of historical reconstruction and that of the social evolutionists. For these latter, the anthropologist's job was the historical reconstruction of a certain period of the general history of humankind. Normally the period in question was a very remote one, and the method was to examine a society that found itself at a stage of cultural evolution corresponding to that specific period. That approach made no sense for

Boas and his followers. For them it was the reconstruction, not of a specific stage of the general history of humankind (given that there was no such a thing), but of the particular history of a particular people.

What about culture as *explanans*, then? For Boas the anthropologist's job was not only to explain culture but also to use culture to explain human behaviour. Boas defined culture as the historically constructed set of traditions, beliefs and customs that we learn as members of a particular society. He argued that a culture plays a crucial role in the configuration of an individual's mind, including both cognition and perception. In this sense, it influences not only our ways of thought but also our perception of the world.

It is interesting to see how Boas reached such a conclusion. He was a native German speaker and he had close relationships with German speaking students at the University of Columbia (New York), where he was teaching. According to Boas, American anthropologists had a duty to document Native American languages. There were no written texts in those languages and they were rapidly disappearing and being replaced by English. Boas thought that this linguistic information might be lost forever unless someone studied and recorded the grammars and vocabularies while native speakers still remained. He asked his students to go to the Indian reservations and learn the natives' languages. Then he came across a rather curious fact. He realised that when his students transcribed indigenous languages phonetically, their transcripts varied according to the student's native language (German, English, Italian, Polish, etc.). It appeared as if depending on one's native language one could hear different sounds, or one would be more sensitive to certain kind of sounds than to others. He generalised this observation to the whole culture.

Maybe culture similarly influences our perception of the world. It is no wonder that the Eskimos have multiple words to refer to snow, as Boas himself observed, as if their culture made them 'see' more 'snows' where we see only one. So how does culture determine human behaviour, according to Boas? By configuring our minds and thereby determining our perception of the world. Where do we find a people's culture then? In their minds. It is also partially embodied in material objects, such as art works, tools and buildings. But these embodiments originate in the makers' minds. Therefore, anthropologists need to be mind explorers. And in order to explore anyone's mind you've got to get him or her talk as much as possible and write down everything. For Boas, this is how anthropologists should conduct their research.

Boasians, and historicists in general, understood that cultures should be seen as accumulations of history in the minds of humans. Tradition is another word that encapsulates this equivalence between culture and particularised history. A tradition is something we obtain from our ancestors and we pass down, with some modifications perhaps, to our descendants. Notice that from this perspective traditions might be totally useless to the individuals who hold them. A tradition can certainly do many different things, such as represent a people's identity in the performance of a ritual. Even more practical concerns can be equally met with traditional knowledge, such as food recipes or agricultural techniques. But that is not what makes it a 'tradition'. What turns any idea or behaviour into a tradition is simply that it has been handed down to us by our predecessors, and we feel obliged to do likewise and pass it down to our successors.

To see cultures as traditions means that humans are useful to cultures rather than the other way around. By following a particular tradition and passing it down to our descendants, humans show how 'faithful' they are to that tradition and how eager they are to keep it even if that entails some sacrifice. Yet other authors argue that viewing human cultures as mere traditions overlooks a very important characteristic of those cultures; namely, that they have been very useful to humans, and not only for expressing collective identities. True, the fact that a tradition turns out to be useful does not make it a tradition. But that utility tells us a very important characteristic of cultural knowledge. We have already seen the effectiveness of cultural knowledge as a means of adaption in the process of human evolution. No human individual could survive a single day without cultural knowledge. Cultures are useful to humans, they meet human needs in a way unparalleled by any other product of biological evolution. Functionalism is the theoretical approach in anthropology that precisely tries to explain culture by focusing on its utility. People are not simply blind followers of traditions; rather, traditions serve functions and the functions themselves explain the existence of the traditions.

Anthropology against history: functionalism

The second milestone in the history of anthropological research is provided by another expatriate, Polish anthropologist Bronislaw Malinowski (1884–1942), who made his career in Britain. Like his German-American colleague, he was a physicist by training, but unlike Boas he did not become interested in anthropology while conducting research among indigenous populations. Malinowski discovered anthropology in books that he read while in hospital in his native Poland. He became so fascinated with the discipline that he decided to emigrate to England to study it. The year was 1910.

Britain was a good place to study anthropology. That was the time when the British Empire spread its tentacles practically all over the world. Myriad different peoples and cultures were under the British imperial administration. Therefore, authorities held a keen interest in studying those peoples and cultures, even if it was just for political and administrative reasons. At the time, British anthropology was still under the intellectual sway of the social evolutionists. Despite evolutionist anthropologists' lack of interest in conducting research directly among the people they studied, some leading British anthropologists had started to collect their information by making short research trips. They would hire an interpreter and ask as many questions as possible to the locals about whatever topic they were interested in. This looks very similar to the sort of research Boas and his students were doing in America. But the differences between the research conducted by British anthropologists and the Boasians were also noticeable. Whereas Boas and his followers emphasised the importance of conducting this type of field research in order to obtain the appropriate kind of data, their British counterparts had a more instrumental view of fieldwork. For them, it was just one means of obtaining information about indigenous societies, but they did not place any particular importance in it.

Malinowski intended to conduct a short field study when he embarked on a trip to Australia in 1914. But something totally unexpected happened while he was there: World War I broke out in Europe. Malinowski found himself in an uneasy situation. He was a Polish citizen so he could not go back to Britain since he was considered an 'enemy' citizen. But returning to Poland would result in him being drafted and sent to war. British colonial authorities in Australia told him that he might as well remain there while there was war in Europe.

Malinowski decided to settle in a small archipelago east of the island of Papua New Guinea called the Trobriand Islands and wait until the war in Europe ended. It was a long wait: four years. In that time, Malinowski conducted a form of research among the inhabitants of those islands that no one had done before. He learned the local language and talked to the natives without interpreters. Not only that. He set up his tent in the middle of the village and lived among the villagers, interacting with them on a daily basis, observing them and participating with them in their daily routines, sharing their lives and experiences. As a result, his information about the people's way of life and ways of thinking was extremely rich and detailed.

Again, we can see some similarities between Malinowski's research and the method employed by Boas and his students in their research among Native Americans. But there were interesting differences as well. Remember the simultaneously historical and mentalistic concept of culture advocated by Boas and his followers. Cultures are historical products that dwell in the minds of the people who possess them. So to study the culture of a society means to study the minds of its members by the only way you can study anyone's mind: by talking to them. Of course, if you could do some participant observation so much the better. But that did not seem to be a *sine qua non* condition for conducting good Boasian fieldwork research. When comparing Boas and Malinowski's methods it is instructive to take into account the different sorts of societies they were studying. Boas and his followers were mostly conducting research on Indian reservations. There was not much to 'observe' in those places since the natives' traditional ways of life had mostly disappeared. Those ways of life were almost exclusively in people's memories. So it is no wonder that those memories became the main object of the anthropologists' research. For Malinowski, and for the British anthropologists who followed his lead, the situation was substantially different. Even though all societies they were looking at were under British colonial rule, they were still very active societies with their traditional ways of life almost intact. So, naturally enough, there was no need to rely on people's memories to study those ways of life when they could be observed '*in vivo*'.

Whatever the case, Malinowski's approach to research has become the standard method for all anthropologists ever since. This is how the ethnographic method and the technique of participant observation came into existence. Following Malinowski's principles, ethnographers conduct research in people's natural settings. They live and develop relationships with them in order to observe them in their daily lives and document their behaviour. Notice that anthropologists conduct their research by themselves becoming the main research tool. But Malinowski's approach to anthropological research was not only a method for gathering information on human diversity, it was also a theory that was meant to explain that diversity.

American anthropologists who followed Boas's lead understood that culture was a fundamental determinant of human behaviour, a result of the way culture organises people's minds. Anthropologists were not oblivious to an underlying common human nature, that the contents of our minds have a lot in common with those of other members of the human species irrespective of the culture in which we have been brought up. But somehow they decided to ignore or downplay this common nature and emphasise the differences. And when it comes to explaining culture itself – culture as *explanandum* as we termed it above – human cultures were seen as contingent products of history. Why does this particular people have this tradition and not the other? This question can only have a historical explanation. They recognised the usefulness of cultures to humans, but thought that utility alone could not explain their existence. As the British economist Kenneth Boulding once said, 'Everything is what it is because it got that way'. Anthropologists who followed Malinowski, by contrast, took a different path or some might say, took a step further.

They rejected the idea that human cultures were merely historical precipitates. Remember that Malinowski saw culture and cultural behaviour from a different angle. He learned not just from people telling him about their culture (a culture that only existed in their memories, in the case of Boas' informants), but also from culture '*in vivo*'. For him, cultures were not mere traditions, something we do because our ancestors did it and told us to do. Instead, cultures fulfilled a very important function for human beings. Humans had invented cultures in order to meet their needs. Malinowski's great idea was that rather than humans being useful to cultures, it is cultures that are useful to humans. This view provided a different way of explaining cultural differences that would go beyond mere historical determination. Cultural institutions could be explained in terms of the human need they satisfied. By way of example, let us see how Malinowski accounted for the belief in magic.

Many peoples all over the world share this belief. According to magical thought, invisible, supernatural powers are responsible for everything that happens. They can be understood as subjects, such as powerful spirits and deities, or as objective entities, such as strange energies. In any case, they are mysterious and essentially incomprehensible to human beings. Yet according to magical thinking, we can influence those powers through magic rituals, making them do what we want. Why do peoples all over the world have these odd beliefs?

Cultural evolutionists thought that it was all the result of ignorance. Peoples who believed in magic were called 'primitive', 'savage', or simply 'backward'. They wanted to control the events of nature, but they did not know the laws of nature. They mistakenly thought that by performing those rituals the desired results would occur. Maybe one day there was a coincidence: I wore a red shirt and I had a successful hunt. From then onwards I thought that if I wanted to excel at hunting,

I needed to wear a red shirt. Why do people believe in such nonsense? Because they are simply ignorant, they take a mere historical correlation (something happens before something else) for a causal correlation (what happens first is the cause of what happens afterwards). Historicists, by contrast, did not think that magic was merely the result of ignorance but rather the result of tradition. People perform magic rituals because that is their tradition. Look at the history of that particular belief, where it comes from, people might have borrowed it from another people. According to historicists, history is why particular peoples have the particular traditions that they have and not others. But historicist anthropology would not go any further than that. Malinowski would.

Malinowski thought that like any other cultural institution, magic exists to meet human psychobiological needs, by which he meant simply psychological and biological needs or the needs of the mind and the needs of the body. Those needs were probably universal but the ways in which humans satisfied them were not; these were the cultures humans had created. Malinowski disregarded the explanations of those pre-modern anthropologists that attributed magic belief to plain ignorance. He gave the example of the agrarian magical rites practised by the Trobriand islanders he knew well. Before tilling the field and sowing their seeds, all sorts of magical rites were prescribed in order to ensure a good harvest. But the Trobriand islanders were not ignorant. They had refined agricultural techniques and knowledge that Malinowski did not hesitate to qualify as 'scientific'. And yet they performed all those apparently useless rites. Why? Certainly, it was not the product of ignorance of farming practice.

Malinowski observed that no matter how accurate their agricultural knowledge happened to be, they could never be 100 per cent sure of the success of their work. Lots of unpredictable things could happen, such as thunderstorms or severe droughts that might spoil the harvest. This uncertainty created a deep state of anxiety in people's minds. They performed their rites in order to come to terms with this anxiety. They knew only too well that in order to have a good harvest they needed to sow their fields and work hard, following their accurate knowledge of agriculture. So no way could they be defined as ignorant people. Yet the anxiety they felt in front of the uncertainties of the future had to be somehow addressed. Take an example that might be more familiar to us: a lover who has been abandoned by his fiancée. He rips a photograph of her into pieces, as if he were actually causing injury to the body of the person depicted in the picture. But he does not really believe that he is causing any harm to her. We go to a demonstration against a particular dictator and we burn a flag, or maybe we burn the picture of that dictator. Do we really believe that we are causing harm to that person? Not really. So why do we do it, then? Simply because we relieve our anxiety in this way. Malinowski thought that magic rites fulfilled this need for catharsis, that is, a psychological need.

Perhaps the natives really do believe in the effectiveness of their magic rituals, by contrast with the frustrated lover or the angry demonstrators. But the existence of these mistaken beliefs cannot be explained as merely the result of ignorance. If they did not fulfil some important function, being wrong as they clearly are, they would have disappeared long ago since they are not actually effective. Anthropologists call this a 'functionalist' explanation. Ever since Malinowski, anthropologists have been quarrelling about what needs cultural institutions satisfy. Are they always 'psychobiological' needs? Or does society itself create its own needs? What psychobiological needs are met when we buy bigger and faster cars?

Structural-functionalism

Another leading figure in British social anthropology, Alfred Reginald Radcliffe-Brown (1881–1955) thought that the functionality of culture should be seen not in terms of individual needs but in terms of social needs. The society where we live, or what he termed its 'social structure', generates the needs that must be satisfied by cultural knowledge. This is the branch of functionalism known as 'structural-functionalism'. At the heyday of British anthropology, during the 1940s, 50s and 60s, structural-functionalism became the dominant theoretical school. From the point of view of this school of thought, cultures are useful, they are not a mere bunch of traditions, but they are not useful to humans as individuals, but rather to the societies where they live. Thus, in order to explain culture one first has to know the characteristics of the social structure that generated that culture. Once we know what sort of needs that social structure has, we will be able to find out why it has that particular culture, for cultures are just 'functions' of the '(social) structure' in which they find themselves.

In a way, structural-functionalism was an improvement on Malinowskian bare functionalism. There was a problem in the way Malinowski's functionalism purported to account for the existence of particular cultures. If it was merely a matter of satisfying individual psychobiological needs, it was unclear how these psychobiological needs were generated in the first place. If they were universal (as the term 'psychobiological' suggests), how could the same universal needs be satisfied in so many different ways? By simply saying that different cultures meet the same universal needs, we are not explaining cultural differences but simply explaining them *away*. If magic beliefs and rituals exist in order to assuage humans' anxieties about the future, why don't all humans believe in magic, since all humans are likely to feel equally anxious about the future? If, on the other hand, human psychobiological needs are not universal, if different individuals – perhaps living in different societies and/or different environments – have different psychobiological needs, then we need another theory that explains why those psychobiological needs are so different. The problem is that in this case we have not explained anything at all. We have simply said that human cultures are different because humans are different. Why do some people believe in magic when others don't? Because they are different, they have different psychobiological needs.

No matter how we define those psychobiological needs, we seem to fall into a theoretical cul-de-sac. Clearly, there has to be something *else* that explains the existence of those beliefs, apart from their role in the satisfaction of universal human psychological needs; something that explains why some people are more anxious than others as regards the future and hence who are more likely than others to entertain beliefs in magic. Structural-functionalism seemed to provide a solution to that dilemma. It is not the universal human psychobiology that generates those needs but rather the societies where humans live. Because human societies differ from each other, they generate different needs and different cultures to satisfy those different needs. Anthropologists who had a structural-functionalist upbringing became obsessed with the study of social structures, for it was commonly believed that social structure was the key to understanding cultural differences.

But structural-functionalism's theoretical improvements were more apparent than real when it came to actually explaining the existence of cultural differences. Structural-functionalism gave hundreds of ethnographic studies the appearance of being 'theoretical', that is to say, they were not mere descriptions of particular people's behaviour and beliefs, but they seemed to provide an explanation for those beliefs and behaviour in terms of the characteristics of the society, or the 'social structure', that had presumably generated them. So far so good, but what does it actually mean to say that 'social structure' makes people behave in a particular way? Are not social structures cultural products themselves? We arrive at the circular explanation that societies generate cultures that in turn produce those societies. So people are different because their cultures (or societies or social structures) are different; but nobody knows where those cultural differences do come from – once the reference to psychobiological needs has been discarded. Maybe they are just contingent historical products, as the historicists had argued. In fact, if we substitute the word 'tradition' for 'society', this does not look all that different from the old Boasian historicism: it is humans that are useful to societies (or traditions) and not the other way around.

A synthesis of historicism and functionalism

Let me conclude this section by making some general observations as regards the theories of culture we have seen so far. The Darwinian approach to biological evolution might give us some interesting clues as to how the dichotomy between functionalism and historicism could be accounted for. From a Darwinian point of view, biodiversity is the result, first of all, of adaptation and natural selection. Because living organisms face different adaptive challenges in different environments, they evolve in different ways. But, clearly, the different adaptive challenges posed by different environments is insufficient to account for biological diversity, otherwise the same environment would always give rise to the same species, and only one. Why is it, then, that the same environment with, presumably, the same environmental challenges to all the organisms that live in it, gives rise to different species? For natural selection to be operative, we need something else, in addition to adaptation, we need random mutations. The important thing to be emphasised here is that those mutations are precisely that – random. Now we can explain biodiversity: there are different species not only because of the different adaptive challenges posed by different environments, but also because those different adaptive challenges act upon something which is *already* different, that is, the different organisms that result from random mutations originated in their process of reproduction. Hence we have two different processes at work here: a necessary process, adaptation; and a contingent process, random mutations.

Now the same applies to our account of human cultural diversity. Historicists saw in culture particularity and contingency, whereas functionalists saw necessity. It is as misleading to define culture as a mere cluster of traditions that humans follow blindly, as it is to define it as something that is always useful. Remember that at one stage we defined cultures as viruses that colonise our brains and find there the appropriate conditions for their reproduction. Perhaps a solution is to rephrase Malinowski's theory of psychobiological needs in terms of humans' reproductive fitness. Taking this point of view, cultures are useful to humans simply because they increase their reproductive fitness. We have already seen that culture helps us reproduce more effectively and in a wide range of environments. We can explain the existence of different human cultures by looking at the way they help particular groups of humans live and reproduce in particular environments. So that is the 'function' of cultures.

In what sense does this approach improve upon Malinowski's theory of universal psychobiological needs? For one thing, we have introduced the variable of environment. Human reproductive fitness is still a pre-cultural given, like human psychobiology. But human reproductive fitness is heavily dependent on the characteristics of particular environments, because by definition reproductive fitness for any organism results from its adaptation to a particular environment. Therefore variability is inherent to the concept of reproductive fitness in a way in which it is not to Malinowski's initial account of psychobiological needs. But that is perhaps a relatively minor point, since the environment variable could be easily introduced in an improved version of Malinowski's theory. It could be argued that human psychobiological needs vary according to the environment in which humans find themselves, and that is why humans have so many different ways of satisfying those universal needs. In both cases we would have environmental variability as the key factor in the explanation of cultural difference. The problem is that there is not a one-to-one correspondence between environment and culture, as Boas himself had perceptively noted long ago. Even though it is true that very different environments will very rarely give rise to the same or very similar cultures, the opposite seems to be much more common, very different cultures can thrive in very similar environments. So there has to be something else other than, or in addition to, environmental variability that explains cultural diversity.

We have already seen that cultures can be adaptive *and* maladaptive, since the paths of cultural transmission are not those of biological reproduction. Cultures can go from having a symbiotic relationship with humans to having a parasitic one. In this case, it is no longer humans that fulfil their needs with the help of the cultures they have created but rather the other way around. We can see in this another way of rephrasing the old Boasian insight that culture is, above all, tradition. Isn't a 'tradition' something that reproduces itself by means of humans? That might be the advantage of seeing cultures as metaphorical viruses that colonise human minds. These viruses have their own 'goal' (to reproduce themselves) that quite often coincides with ours (to reproduce ourselves too), *but not always*. In a nutshell, humans reproduce themselves by using cultures in the same way that cultures reproduce themselves by using humans. In this manner we have managed to integrate both the historicist and the functionalist approaches in anthropology in a unified theoretical framework – though with some modifications to their initial formulations. A historicist approach (within which we should include the structural-functional version of functionalism) emphasises the capacity of cultures to reproduce themselves by means of humans. A functionalist approach, by contrast, underscores the capacity of humans to reproduce themselves by means of cultures.

Notice, incidentally, that the fact that cultures use humans to their own ends, as it were, is not necessarily a 'bad thing' in strictly moral terms. Nor is the fact that humans use cultures to meet their own needs necessarily a 'good thing', either. Think of a soldier who sacrifices himself for his country, or a freedom fighter who sacrifices herself for freedom, democracy or justice. 'Freedom', 'democracy', and 'justice' are mere cultural constructs that can be seen as good or bad depending on one's perspective. Some people might be willing to sacrifice themselves for the sake of replicating these cultural constructs, and we might consider them heroes. Similarly, we might not think highly of someone who uses those beautiful ideals to her own benefit (maybe to increase her own personal wealth and, as a result of it, her own reproductive fitness) to the detriment or exclusion of others.

Structuralism and the problem of meaning again

This brief overview of culture theories would not be complete without a reference to structuralism, surely one of the more theoretically sophisticated approaches to cultural difference in classical anthropology. Structural anthropology was born in France and it was the creation of, again, perhaps the most distinguished theoretical anthropologist of the twentieth century: Claude Lévi-Strauss. Despite its undisputable merits, structuralism has remained underdeveloped as a general anthropological theory outside the confines of Lévi-Strauss's outstanding work and that of his most brilliant disciples, both within and outside France. Whatever the reasons for this relative bleakness in current anthropology's theoretical landscape, there are a few aspects of the structuralist project that require our close attention because they are directly relevant to the approach to cultural difference we have been developing in this text. Prominent among them is the question of meaning. For structuralism, cultures are above all meaning systems, and structuralists study *and* explain them as such.

This makes structuralism notably different from both historicism and functionalism. Cultures can neither be seen as merely inveterate traditions that humans learn and pass on to the next generation nor as instruments for the satisfaction of human needs. Cultures certainly have these characteristics, but these are not their distinctive characteristics; they share these characteristics with other forms of knowledge that also govern human behaviour, such as genetic knowledge – which is also inherited – or individual knowledge, which also satisfies human needs. What distinguishes cultures as determinants of human behaviour is their meaning-giving capacity: cultures govern behaviour by making it meaningful.

We have already defined meaning as the other side of the coin of an intention. Just as intentional behaviour originates in my mind (an emergent property of my brain), meaningful behaviour originates in the collective mind that is our culture (an emergent property of interacting minds). True, meaningful behaviour is also intentional, but 'my' individual intentions alone are not responsible for the production of that behaviour. If they were, it would be an intentional but meaningless behaviour. Now the question is: how do we approach the analysis of these meaning systems?

In our day-to-day life we do not normally need to understand the culture that makes other humans' behaviour meaningful, because that culture is part of our upbringing. But anthropologists find ourselves in a different situation. Because we do not know other people's cultures, the behaviour that is meaningful in those cultures is meaningless to us. The anthropologist's job is to learn other people's cultures. To this end, modern anthropologists try to emulate through ethnographic research the process of culture learning that society members go through. Of course this emulation is always very limited because anthropologists cannot re-socialise themselves in the culture under study. But there is a sense in which ethnographic research in the way we have described it can be seen as an attempt at precisely that. In Malinowski's celebrated dictum: we have to see the world 'from the native's point of view'. Once we have reached that stage, that initially alien culture will no longer be a mystery to us, and therefore the behaviour of those who have been brought up in it will become transparent to us: it will become a meaningful behaviour.

Here is where structuralism parts company with the other classical anthropological schools. It might be worth it at this juncture to refer to the well-known contrast between Malinowski's celebrated fieldwork in the Trobriand Islands and the ethnographic research that Lévi-Strauss carried out in the Amazon rain forest in the late 1930s. Whereas the first always took his field research as his most undisputable merit as an anthropologist, Lévi-Strauss never seemed to be fully satisfied with his, as if the practice of fieldwork had left him with a deep sense of impotence and frustration. Even though he always encouraged all his students to conduct long-term ethnographic fieldwork in the Malinowskian way, for him fieldwork in itself could never be seen as coextensive with anthropological research. There had to be more in anthropology than that because ethnography is always, almost by definition, an incomplete project. No matter how deep our understanding of a particular group of people appears to be, we will never be able to fully reproduce their life experiences. And even if we could, we would probably be unable to communicate that knowledge to anyone other than the people themselves. Does that mean that anthropology's project, understanding all humans, is doomed to failure? Not really. It would mean that the understanding aimed at by anthropologists was merely an extension of the sort of understanding that we achieve in ordinary communication with other human beings, normally belonging to our own culture. But that is not the way Lévi-Strauss saw the objective of anthropology.

Anthropologists are not so much interested in revealing the actual meaning of other people's behaviour (an impossible task, as I have just argued), but in the process of producing that meaning; in other words, we don't want to know *what* they mean but *how* they mean. We can try to make the same point by using the old Boasian understanding of culture.

After his stay in the Brazilian Amazon, Lévi-Strauss went to the US to study anthropology with Boas and his followers, and Boasian historical particularism made a remarkable impact on Lévi-Strauss's structural anthropology. Remember that Boas saw human culture as the contingent product of history; as such, it could only have a historical explanation. But this is a rather peculiar way of explaining culture. Why do these people have this particular belief or that particular custom? Simply because they got it from someone else. But, unlike in functionalism, there does not seem to be any intrinsic rationale for the belief or custom as such. In a way, a historical explanation of culture is an admission that cultures cannot be explained – they are contingent. Structural anthropology has often been accused of ignoring the historical nature of culture. But that is a very unjust accusation. In fact, Lévi-Strauss was fully aware of culture's historicity. In the same way as we can never fully grasp the meaning of an alien cultural practice, we cannot explain why particular peoples have the particular cultures they have, for history is a contingent process. However, what we *can* do is to reveal how that culture is produced as a meaning system.

If cultural meanings can be understood as collective intentions, culturally meaningful actions have the same characteristics as intentional acts, notably the fact that an intentional act is one that could have *not* been produced. If my actions are the result of my intentions, in order to change my actions all I need to do is change my intentions. This is what free will is all about. Otherwise stated, the meaning of my intentional action lies in the fact that I could have done the very opposite. The action of reading this book is a meaningful action because it is intentional; namely, instead you could have burned it or thrown it into the bin. So all intentional acts have their opposite as a sort of latent possibility. Similarly, a meaningful action must have its negation as a latent possibility; that is precisely what makes it meaningful. This looks like a pretty obvious assertion, though its consequences are much less obvious. It makes sense to talk about fat people because some people are thin. It makes sense to talk about long roads because some roads are short. In this sense, everything seems to gain meaning from its opposite. Now we will see how to turn this apparently banal statement into a method of cultural analysis.

Imagine that we could arrange all culturally meaningful actions since the beginning of human history (as far as our historical and ethnographic records permit), in a huge classificatory matrix in which each particular action sits alongside its opposite. Sometimes the opposite would be a latent or logical possibility never actualised; other times, it would be an actually existing action, to be found either in a different culture, or in the same culture at a different period of time. Notice that in this huge matrix of human behaviour all cultural actions would become meaningful actions, not because we happen to know very deeply the people who actually performed those acts, but rather because we would be able to compare them with their negation in an endlessly interlocking system of binary oppositions. Believing in one god is a meaningful cultural action or condition because there are (or could have been) people who believed, say, in more than one god. And both believing in one god and believing in several gods gains its meaning because there are or there could have been people who believe in none. That is, we see a second binary possibility of belief vs non-belief. The binary code of this classificatory matrix looks very much like the binary code of computer language. And it also has clear affinities with the systems of binary oppositions that enable the human mind to unconsciously assimilate fundamental categories of language. As structural linguistics shows us, language is based on a set of binaries, for example at the phonemic level (consonants/vowels, fricatives/affricates, voiced/voiceless...), the syntactic level (singular/plural, subject/predicate, noun/verb...) or at any other level. The pervasiveness of binary codes in the products of the human mind cannot be a coincidence. This is how minds actually work. This is the *how* of meaning production.

Does this sound too far-fetched? It probably is. Lévi-Strauss never fully developed the manifold implications of his theory of meaning and the human mind. So, the apparent analogies between binary codes at different levels of mental processing might turn out to be just that, superficial similarities with no substantial affinity. Be that as it may, it is worth taking a quick look at how Lévi-Strauss himself tried to demonstrate the virtues of structural anthropology in his first major work, *The Elementary Structures of Kinship* (first French edition published in 1949), one of the most accomplished works of anthropological structural analysis ever produced.

The problem Lévi-Strauss sets out to solve is that of the diversity of marriage systems. Why have human societies devised so many different ways of organising marriage? Anthropologists had been accumulating information about marriage systems in different societies for a very long time, but nobody knew why those marriage systems were so different. In any human society, marriage always appears to do the same thing: regulate sexual relationships so that the biological reproduction of the group can proceed in socially accepted ways. A functionalist approach was clearly insufficient, not so much to explain marriage as such, but rather to account for the diversity of marriage systems. If they all fulfil the same function why are they so different? Is this merely the result of historical contingency? Lévi-Strauss starts by redefining the function of marriage: it is not just a matter of regulating sexual relationships in socially accepted ways. That is how we tend to see it in our own society, but anthropology must take a species perspective. For the majority of human societies, and certainly for the type of societies wherein humans have lived for most of their history, marriage was not just about sex regulation. In fact, the regulation of sexual behaviour had to be understood within a wider framework of social relations. Lévi-Strauss argued that marriage was at its core a system for regulating the exchange of women between groups of men.

That sounds really odd to Western ears. For starters, many take offence at the notion of men 'exchanging' women. And even leaving that point aside for now, why would men exchange women among themselves? According to Lévi-Strauss, exchange is what constitutes human societies. To live in a society actually means just that: to exchange things, valuable things, with other people. Now for much of human history (that is, since before the Neolithic revolution), women have been the primary form of wealth for any human society. In industrial and post-industrial societies we tend to associate wealth with material goods. But this is not how pre-industrial (and *a fortiori* pre-agricultural) societies would look at it. In such societies few material goods are necessary for survival, and more importantly, they are easy to produce. Lévi-Strauss argues that women are the wealth of such societies. To put it in rather crude terms, as any stockbreeder knows only too well, female animals and not males, constitute the wealth of a herd.

But if the females of the group are so important, why would the men be interested in exchanging them for other females? The females of any group are likely to be the daughters and sisters of the men. Because of the incest taboo, men have to exchange them for other females because they cannot have sex with them. The incest taboo is simply the rule that forbids sex between close kin, normally between parents and children and between brothers and sisters. This seems to be a universal prohibition. All known human societies forbid those sexual relations (with some qualifications that need not detain us now), normally placing severe sanctions on those who break that prohibition. Lévi-Strauss placed great emphasis on the emergence of the incest taboo in human history as the boundary that separates, in his terms, nature from culture or the non-human from the human. Following many authors before him, most prominently Freud, Lévi-Strauss understood the incest taboo as the first cultural norm in human history, what turned our pre-human ancestors into cultured humans. From then onwards human behaviour was no longer ruled only by instincts but by instincts *and* cultural norms, the first of which was precisely the incest taboo.

This is a highly controversial statement. Certainly, non-human animals do not have an incest taboo but they do have incest avoidance. And that is for plain evolutionary reasons: incestuous relationships are likely to increase the rate of recessive hereditary diseases. So natural selection has devised several mechanisms to prevent closely related individuals from mating each other. Lack of sexual attraction between individuals who have been brought up in close proximity would be one of them. This is what is known as the 'Westermarck effect', in honour of the Finnish anthropologist Edvard Westermarck who was the first to hypothesise it. The Westermarck effect amounts to incest avoidance, for those who have been brought up in close proximity are likely to be parents and children or brothers and sisters.

It is true that incest avoidance, an instinctual drive, is not the same as the incest taboo, a culturally constructed social rule. But if incest avoidance exists among non-human animals, and among humans too, it is unclear in what sense the incest taboo sets this stark boundary between the human and the non-human. Let us keep this in mind and proceed with our exploration of Lévi-Strauss's argument. For

Lévi-Strauss the appearance of the incest taboo is of crucial importance because the distinction he sets up between nature and culture marks the birth of meaning. By obeying the incest prohibition and not having sex with each other, a brother and a sister turn sexual behaviour into an intentional behaviour (not merely instinctual). But notice that this is not an intention that originates in their minds, but in the collective mind of the society that imposes this prohibition. Thus, having sex with a socially accepted partner becomes an intentional-meaningful behaviour precisely because one could do otherwise. Consequently, the distinction between marriageable and non-marriageable becomes the first binary opposition in the classificatory matrix of human acts that should theoretically encompass the whole of human culture.

From here Lévi-Strauss unpacks the whole system of binary oppositions. There is no need to go into too much detail here but it is worth taking a cursory look at Lévi-Strauss' argument. After the initial distinction that the incest taboo establishes between what we might call human and non-human mating systems, these latter can, in turn, be divided into elementary and complex, according to whether they have a prescriptive marriage rule in addition to the incest prohibition. Complex systems do not have a prescriptive rule; people living in such systems can marry whomever they wish except those affected by the incest prohibitions. These are the systems currently prevalent in Western societies and a few others. Elementary systems, by contrast, express the marriage rule in positive terms, identifying not the class of people whom one should avoid marrying but the class(es) of people whom one *should* marry. Elementary marriage systems are to be found in the majority of tribal societies and they generate what Lévi-Strauss calls 'Elementary structures of kinship', the title of his 1949 book.

The origins of this distinction between prescriptive and non-prescriptive marriage rules is unclear. But from Lévi-Strauss's overall argument we might infer that it has to do with changes in the relative value of women as objects of exchange. By exchanging women, men not only can have sex (and therefore reproduce), without violating the incest taboo, but also they can also form alliances among themselves (and in so doing, constitute society). As societies became capable of producing other valuables such as material goods, and these become even more valuable than women themselves, the exchange of these goods came to constitute the fundamental structure of the society. Therefore we must assume that there is no need for a prescriptive rule then. This can be clearly seen in the evolution of marriage transactions, which go from direct sister exchange, wherein men exchange directly their sisters with each other (for nothing can be given in exchange for a woman other than another woman), to different forms of bridewealth, in which women are given in exchange not for other women directly but for something that 'stands for' a woman, such as cattle or material goods. Dowry payments would be the final stage in this evolutionary sequence: now women have become so 'valueless' compared to material goods that women (or their families) have to pay to get married instead of the other way around. Whereas dowry payments appear in complex marriage systems, bridewealth and direct sister exchange are characteristic of elementary systems.

The core of Lévi-Strauss's analysis is the study of elementary systems, which are in turn divided into two kinds: restricted exchange and generalised exchange. In the first, the men of group A and group B exchange their sisters directly; in the second, men from group A give their sisters to men from group B and obtain wives from group C. Normally, men from group A receive from group B something that stands for a wife: the bridewealth payments that they will subsequently use to obtain wives from group C. So a generalised exchange system would correspond with a society capable of producing some form of material wealth commensurable with women. As for the restricted exchange systems, the simplest is the moiety system, in which society is divided into two halves and the marriage rule simply prescribes marriage with someone belonging to the other half. After that we have another system slightly more complex, the section system, in which each half is in turn divided into two further halves, called 'sections', or marriage classes. The marriage rule prescribes marriage into one of the four sections. The section system is followed by the sub-section system, with each section being divided into two and thus giving rise to eight sub-sections. In theory, the process could proceed indefinitely by dividing each marriage class and giving rise to systems with 16, 32, 64, etc. marriage classes. But ethnographic evidence, mostly coming from Australian aborigines, only goes as far as the sub-section system; the others remain pure logical possibilities that have never been actualised. It is not difficult to see why beyond the sub-section system those unrealised logical possibilities become sociologically unfeasible, for in each system there has to be a predetermined number of marriage groups that rigidly follows the formula 2^n, where n stands for an ordinal number that refers to each marriage system: moieties $n=1$, sections $n=2$, sub-sections $n=3$, etc. But human societies have devised the opposition restricted/generalised exchange precisely to get over that limitation, for in generalised exchange there is no formula determining the exact number of marriage groups, other than the requisite that there have to be more than two since, as we have seen, any group of men must obtain their wives from a different group from the one that has received their sisters.

Whether Lévi-Strauss' detailed analyses of marriage systems can really be extended to all cultural phenomena is a moot point. For our purposes, the important point is that in structuralism, each particular marriage system is exclusively explained not in terms of its function or its history, but in terms of its meaning. To repeat the argument that has already been made: behaviour within those marriage systems becomes meaningful in so far as it can be placed alongside its very opposite in a system of binary oppositions. What makes any human action meaningful is that it could have been its very opposite. Now the opposite of each marriage system, and the behaviour it regulates, is another marriage system, and this is a marriage system that can actually be found (most of them) in the ethnographic record. But this is not the meaning that could be located in people's heads and disclosed through Malinowskian ethnographic practice. People do not have to be conscious of the rules of the marriage system in which they live; rather this information merely has to be in their unconscious. And by unconscious, Lévi-Strauss understood the deepest structures of the human mind that we all share and about which we know nothing, despite the fact that they actually govern our behaviour, in the same way as the grammar of our language governs our speech unbeknown to us.

Beyond structuralism

Let us go back now to the issue of the contingency of culture as exemplified by the Boasian approach. Cultures are different not because they fulfil different functions or the same function in different environments, but fundamentally because they are the contingent product of human history, equivalent, in this sense, to the random mutations of DNA sequences in biological reproduction. What do we do with something 'contingent'? Structuralism could be defined as an attempt at coming to terms with this contingent nature of culture. We can write the history of contingent facts, certainly, but there does not seem to be any way in which we can explain those contingent facts in functional terms, for contingent means precisely that, something that cannot be explained or accounted for. In reality, a contingent fact is a fact that could have been otherwise. In structuralism we see an approach to the contingent nature of cultural facts from within, as it were, by facing head-on the fundamental characteristic of culture as an emergent property of interacting minds: its capacity to produce meaning. And in structural analysis, as we have seen, a meaningful action is an action that could have been otherwise. In functionalism, by contrast, we see an approach to culture from without, that is, we want to find out what cultures are there for, whether this be the satisfaction of human needs or the adaptation to particular environments. Hence cultural facts stop being contingent when we look at them from without, for they exist because they fulfil some purpose. Both approaches are useful and informative as long as we do not forget the different perspective each one takes on the same object, or on different dimensions of the same object. We shall return to this important distinction at the end of the next chapter.

Bibliographical note on Chapter 3

As regards the ethnographic method of research, a good account can be found in Hammersley and Atkinson (2007), and see also Agar (2008) for a more relaxed, though equally rigorous, overview. The research techniques of participant observation and the ethnographic interview are thoroughly examined in Spradley's classical handbooks (1979 and 1980). Both are practically oriented texts with, perhaps, an excessive formalisation. More theoretically informed, though dealing exclusively with the analysis of interviews and other forms of discourse, is the work edited by Naomi Quinn (2005). In fact, there is no fixed set of rules for conducting participant observation research. Participant observation is a form of living and no predetermined regulation can tell anyone how to live. The best option is, in my view, to study critically first-person accounts of the experience of fieldwork, such as Malinowski's famous diary (1989), and the excellent reanalyses and historical contextualisation of Stocking (1985). More critical reflections on the practice of fieldwork became quite fashionable in the 1980s and 1990s. An interesting and early instance of this critical mood can be found in Rabinow (1977). In what concerns the history of anthropological theories, Barnard (2000), Layton (1997) and Salzman (2001) provide good and comprehensive introductions. Kuper (1988) contains a more specialised text, which is less comprehensive and more critical and argumentative. See also Kuper (1999) for a critical history of the concept of culture centred on the American anthropological tradition and Kuper (2015) for an updated history of the British school of anthropology. For a deeper and detailed historical analysis of the development of classical anthropological theories see Stocking (1982, 1985, 1998). Needless to say, the reader who may wish to go into the original texts written by representative authors of each school has abundant material at his or her disposal. For cultural evolutionism, the classical texts by Tylor (1920) and Frazer (1923) are still highly readable, especially for those keen on Victorian English prose. Boas's fundamental texts can be found in Boas (1995). See also Boas (1938) for a concise statement of his main ideas and Lowie (1917) for a classical defence of the historicist approach to anthropology. Malinowski's groundbreaking research on the Trobriand islanders (1922) is undoubtedly the best exponent of a functionalist ethnography. More theoretical works on functionalist anthropology are Malinowski (1944, 1948). The best representative of the structural-functionalist theory in its most classical formulation is Radcliffe-Brown (1952), a finely argued defence of the principles of sociological functionalism that became the gospel of the British school of anthropology. For more ethnographically-oriented approaches to structural-functionalism see Evans-Pritchard (1962) and the monographs by this same author on the Azande (1976) and the Nuer (1940). There are interesting theoretical differences in the way in which sociological functionalism was developed by Radcliffe-Brown and Evans-Pritchard. Whereas the first defined anthropology as the 'natural science of society', the second saw anthropology as a humanistic rather than scientific discipline. Finally, in what concerns French structuralism, the outstanding work of Lévi-Strauss has been the object of numerous analysis and interpretations. See, for instance, the classical work by Leach (1970) and the more recent publications by Deliège (2004) and Wilcken (2010). Despite the fact that Lévi-Strauss was an excellent writer (or maybe because of that) his texts are not easy reading for beginners, and the English translations have not always been very accurate. Even though Lévi-Strauss never published a full and systematic theoretical account of the principles of structural anthropology, the two volumes entitled 'Structural anthropology' (1963 and 1983) constitute the best approximation. Perhaps the most readable exposition of Lévi-Strauss' thought can be found in the set of interviews conducted by the French philosopher Didier Eribon (1991). Be that as it may, the inexcusable text for anyone interested not only in Lévi-Strauss or French structuralism but also in the making of modern anthropology is *The Elementary Structures of Kinship* (1969). This is unquestionably a masterpiece of the social sciences of the twentieth century. Erudite, theoretically dense and cleverly argued, it is still by far the finest example of the structuralist analysis of culture.

References

Agar, M.H. 2008. *The Professional Stranger*. Bingley: Emerald Group.
Barnard, A. 2000. *History and Theory in Anthropology*. Cambridge: Cambridge University Press.
Boas, F. 1938. *The Mind of Primitive Man*. New York: Macmillan.
———. 1995. *Race, Language, and Culture*. Chicago: University of Chicago Press.
Deliège, R. 2004. *Lévi-Strauss Today*. New York: Bloomsbury Academic.
Eribon, D. 1991. *Conversations with Claude Lévi-Strauss*. Chicago: University of Chicago Press.
Evans-Pritchard, E.E. 1940. *The Nuer*. Oxford: Clarendon Press.
———. 1962. *Essays in Social Anthropology*. London: Faber and Faber.
———. 1976. *Witchcraft, Oracles and Magic among the Azande*. Oxford: Oxford University Press.
Frazer, J. 1923. *The Golden Bough*. London: Macmillan.
Hammersley, M. and P. Atkinson. 2007. *Ethnography. Principles in Practice*. London: Routledge.
Kuper, A. 1988. *The Invention of Primitive Society*. London: Routledge.
———. 1999. *Culture: The Anthropologists' Account*. Cambridge, MA: Harvard University Press.
———. 2015. *Anthropology and Anthropologists: The British School in the Twentieth Century*. New York: Routledge.
Layton, R. 1997. *An Introduction to Theory in Anthropology*. Cambridge: Cambridge University Press.
Leach, E. 1970. *Lévi-Strauss*. London: Fontana Press.
Lévi-Strauss, C. 1963. *Structural Anthropology, Volume 1*. New York: Basic Books.
———. 1969. *The Elementary Structures of Kinship*. Boston, MA: Beacon Press.
———. 1983. *Structural Anthropology. Volume II*. Chicago: University of Chicago Press.
Lowie, K. 1917. *Culture and Ethnology*. New York: Boni and Liveright.
Malinowski, B. 1922. *Argonauts of the Western Pacific*. London: Routledge and Kegan Paul.
———. 1944. *A Scientific Theory of Culture*. Chapel Hill, NC: University of North Carolina Press.
———. 1948. *Magic, Science, and Religion*. Glencoe, IL: Free Press.
———. 1989. *A Diary in the Strict Sense of the Term*. Stanford, CA: Stanford University Press.
Quinn, N., ed. 2005. *Finding Culture in Talk*. New York: Palgrave Macmillan.
Rabinow, P. 1977. *Reflections on Fieldwork in Morocco*. Berkeley, CA: University of California Press.
Radcliffe-Brown, A.R. 1952. *Structure and Function in Primitive Society*. Glencoe, IL: Free Press.
Salzman, C.P. 2001. *Understanding Cultures*. Long Grove, IL: Waveland Press.
Spradley, M.H. 1979. *The Ethnographic Interview*. Belmont, CA: Wadsworth.
———. 1980. *Participant Observation*. Belmont, CA: Wadsworth.
Stocking, G.W. 1982. *Race, Culture, and Evolution*. Chicago: University of Chicago Press.
———. ed. 1985. *Observers Observed*. Madison, WI: University of Wisconsin Press.
———. 1998. *After Tylor: British Social Anthropology, 1888–1951*. Madison, WI: University of Wisconsin Press.
Tylor, E.B. 1920 (1871). *Primitive Culture*. London: John Murray.
Wilcken, P. 2010. *Claude Lévi-Strauss. The Poet in the Laboratory*. London: Penguin Press.

4
CULTURAL EVOLUTION

The puzzle of cultural change

In structuralism we have seen a theory of synchronic cultural diversity, that is, cultural diversity in space. The fact that some of the cultural forms that can be plotted in the structualist classificatory matrix happen to be past cultural forms is irrelevant to their significance, for time does not seem to play any role in explaining cultural difference. From another point of view, it could be argued that the temporal variable was not a proper object of knowledge for structuralism, in so far as it provided a view of cultures from within. Cultural change could be seen from this perspective as something external to cultures themselves and, therefore, beyond the jurisdiction of the structural analysis of culture. But the time variable was absent not only in structuralism; with the exception of nineteenth century social evolutionists, diachronic cultural change was disregarded by all classical approaches. True, historicists saw change in cultures, but that change was contingent and, therefore, unaccountable. In traditional functionalism, on the other hand, cultural change was never properly analysed, for it was unclear how something that fulfils a particular function should change in any way once that function has been fully satisfied.

However, the truth of the matter is that cultural diversity can also be diachronic. Cultures vary in time as well as in space. It is undeniable that cultures do change and that these changes are not totally contingent. Furthermore, not all cultures change at the same rate. Some cultures seem to be extremely stable whereas others change very fast. How could we explain this variation in the durability of cultural practices? Does it have to do with the fact that some are adaptive while others happen to be maladaptive? Adaptive cultures are useful to humans in the same way that humans are useful for the reproduction of maladaptive cultures. But the adaptive/maladaptive feature is not the only qualification we can place upon culture when we look at the relationship between culture and human biology.

In Chapter 2 we introduced a further distinction; that between intuitive and counterintuitive cultures. Cultures can be adaptive or maladaptive and intuitive or counterintuitive. We can sort them into 2x2 matrix that generates four possible combinations: adaptive intuitive, adaptive counterintuitive, maladaptive intuitive and maladaptive counterintuitive. From this, it would seem reasonable to conclude that counterintuitive cultures are the most unstable, be these adaptive or maladaptive; that is to say, it is the quality of being intuitive or counterintuitive, and not that of being adaptive or maladaptive which explains the stability or instability of cultural practices. But how could it be possible that the fact of being adaptive or maladaptive has nothing to do with the stability of a cultural practice, taking into consideration that cultural adaptation has been one of the most remarkable characteristics of the human species? Durability of cultural practices seems to be related with the extent to which they are intuitive or counterintuitive, no doubt about that, but also with the fact of being adaptive or maladaptive. However, it is not obvious how we can relate adaptation or maladaptation to stability and instability, and much less how those two different sets of characteristics interact with each other.

The theory (or theories) of cultural evolution is supposed to provide an explanation of the success of a particular cultural practice or cultural belief in human history. A brief review will help us to better grasp how the notion of cultural evolution began to gain purchase again on the human sciences. We saw in Chapter 3 that after the demise of nineteenth century social evolutionism, anthropologists became very sceptical about the possibility that cultures could 'evolve' in any way. The problem with that time-honoured version of social evolutionism is that it was hopelessly value-laden (or so it seemed). Cultural practices had to be sorted out into 'more' or 'less' evolved, and there was no clear and objective criterion for doing so – other than by simply considering whatever looked like Western culture as being more evolved and vice versa, the less Western-like the more primitive. After the demise of social evolutionism, cultural relativism became the norm in anthropology: human cultures are incommensurable; they are unique in their own ways, so no one can be seen as superior or inferior, 'more evolved' or 'less evolved'. Note that this form of cultural relativism applied equally well to both the historicist, functionalist and the structuralist paradigms. All cultures are equal, either because they all satisfy human needs or because they are all the product of contingent historical processes.

But from another point of view something seems to be amiss with this cultural relativism, especially when the comparison between cultural traits takes a historical or temporal dimension. It is appropriate to see as equal all cultures that coexist at a particular point in time. In fact, that is what Darwinian evolution posits in regard to biological species. But Darwinian evolutionists also tell us that any of those co-present, equally evolved, biological species come from antecedent forms that can only be seen as less evolved versions of their successors. What accounts for the difference between the more or less evolved is the manifestly objective fact of being more or less adapted to a particular environment. Since adaptation is the result of natural selection, which happens to be a relatively slow process, each species can only reach this valuable objective after a long and painstaking succession of less adaptive versions of itself. So in biological evolution, there are certainly more and less adapted species, but only if we take a diachronic point of view, namely only when we compare any one particular species with its ancestors. Now could a similar process be observed in the history of human cultures? In other words, is human history really a contingent process, as the historicists claimed, or is there still some 'rationale' behind it, once the anti-scientific and value-laden theories of the first social evolutionists have been rightfully and definitively discarded?

Certainly, the idea that cultures could be usefully compared with viruses of the mind does not in itself contradict the idea of cultural evolution. Far from it, are not viruses themselves the result of an evolutionary process? So the same should apply to cultures. Or are we pushing the analogy between viruses and cultural traits a bit too far? Well-known British biologist Richard Dawkins espoused a version of this epidemiological model of cultural transmission, the so-called memeticist theory of culture. This theory goes so far as to postulate the existence of cultural inheritance units, memes, which play the same role in cultural evolution as genes do in biological evolution (therefore the deliberate resemblance between the two words 'meme' and 'gene'). We shall return to this. But first let me go back to the more basic notion of cultural evolution. To repeat: do cultures evolve in any way or is human history an utterly contingent succession of cultural forms?

'Utterly contingent' is clearly too strong a qualification. In human history we do not see societies move arbitrarily from the industrial revolution to the hunter-gatherer mode of production, and then to subsistence farming; these systems tend to appear historically in a given order. No ironclad law of evolution determines the history of economic systems with the same precision as the laws of physics determine the movement of the planets, or perhaps natural selection determines the evolution of biological species. But clearly the history of economic systems is not 'utterly contingent'. Likewise, cultural forms do not tend to emerge or disappear out of the blue, or change into their opposites. But cultures do definitely change, pretty much like biological species. What, if anything, governs those changes? None of the classical theoretical approaches in the history of anthropology that came after social evolutionism (historicism, functionalism or structuralism) was able to provide an answer to that question. For historicism and structuralism, cultural change was merely contingent. For functionalism cultural change was simply unthinkable. Functionalists saw logic in culture, the satisfaction of human needs, but they did not see any logic in cultural *change*, for once a particular culture has met those human needs, there does not seem to be any reason to change it.

If we take a mere superficial look at the process of cultural reproduction we can easily see why human cultural history can never be an utterly contingent process. Newborns do not invent or produce their cultural knowledge anew at each generational replacement; rather, they inherit most of it from their elders. Interestingly, an intriguing parallel between cultural and biological reproduction seems to insinuate itself. Except for identical twins, we are all genetically unique but we all receive the totality of our genes from the members of the previous generation (our biological parents). Besides, the genetic differences between any two members of the same species are certainly very few – that is what makes them qualify as being of the same species. Thus, making allowances for random mutations and for the genes that could not get themselves copied, in any particular species the whole genetic endowment of each new generation is identical (if recombined) with that of the previous generation. Evolution results from the effects that natural selection has upon this new recombined set of genes plus the few mutations that we find in each new generation (which are normally negligible or inexistent in the short run but quite remarkable in the long run). Could a similar process take place in cultural reproduction? Cultures do seem to go from one generation to the next, the same as genes. They also accumulate changes a little at a time, like genes do. How far can we push the analogy between genetic and cultural inheritance?

Evolution of cultures revisited

It is hardly surprising that the first authors who, back in the 1980s, started to wonder whether evolution occurred in human cultural history were neither anthropologists nor social scientists but natural scientists. These were the geneticist Luigi Cavalli-Sforza and the mathematician Marc Feldman of Stanford University, and ecologist Robert Boyd and biologist Peter Richerson of the University of California. In an introductory text like this we will not go into detail on the seminal theories of cultural evolution that they put forward, incidentally, lavishly seasoned with rather sophisticated mathematical models, which certainly contributed to keeping most social scientists at bay for quite a while. A few general observations will suffice. First of all, even though they all accepted the need to apply a Darwinian evolutionary model to cultural history, they also conceded that such a model had to be substantially different from that of evolutionary biology. The key question is how cultural and biological evolution are the same and how they are different.

On the one hand, we have the process of natural selection, which, according to those founding fathers of cultural Darwinism, would operate in a very similar way in cultural evolution as in biological evolution. In virtue of this process, maladaptive cultural forms will be selected against and the adaptive, or more adaptive, ones will be selected for. But adaptive or maladaptive for whom? Adaptive or maladaptive for human individuals as biological organisms. In other words, cultural practices that enable humans who adopt them to (biologically) outreproduce those who don't will spread. By the same token, cultural practices that do not enable enhanced reproduction will disappear.

But for those early (modern) cultural evolutionists natural selection was clearly insufficient to account for the whole process of cultural evolution, for all human societies contain maladaptive cultural traits that reproduce themselves regardless of their biologically maladaptive qualities. So something else besides natural selection is going on in cultural evolution, something that we might call 'cultural selection' – even though not everybody would have accepted this term. By cultural selection I mean the process of differential adoption and transmission of cultural traits, as the

British psychologist and cultural evolution theorist Alex Mesoudi has recently put it, and which does not seem to follow the logic of natural selection. Cavalli-Sforza, Robert Boyd and colleagues devoted much of their work to the study of cultural selection; that is to say, to the different mechanisms that increased the possibility that a given cultural trait could replicate and that had nothing to do with natural selection, that is, irrespective of it being adaptive or maladaptive. This is what they identified as 'biases', which could, in turn, be classified into three main types: content biases, model-based biases and frequency-dependent biases.

Content biases determined that a cultural trait would be selected according to its intrinsic qualities, whatever these happen to be. Model-based biases, by contrast, accounted for a selection of cultural traits that had nothing to do with their qualities but rather with the prestige of the individual who had them. This is a well-known technique of cultural reproduction in the advertisement industry: ads show famous individuals using a particular product, on the assumption that the mere fact of seeing that individual wearing those trousers or drinking that drink will automatically increase sales. Finally, frequency-dependent biases make cultural reproduction dependent on the raw number of people who adopt a particular cultural trait. The underlying assumption here is that people will tend to follow the majority, regardless of the usefulness of the behaviour.

In addition to these selective mechanisms or biases, they also identified a process through which cultural traits could be modified during transmission, what they termed 'guided variation'. Biologists would probably categorise this process as a 'Lamarckian inheritance', that is, the inheritance of acquired characteristics. Guided variation means that we do not merely adopt a particular cultural trait and pass it on to the next generation, but that we actually change that trait, maybe because it disagrees with our individual knowledge. A more radical version of this guided variation would be 'cultural mutation', in which a completely new cultural practice is thrown into the cultural-gene pool of a particular society. Furthermore, forms of cultural transmission were also supposed to affect the possibilities of replication of cultural traits. These could be vertical, oblique and horizontal, according to whether cultural knowledge came from parents, from other adults belonging to the parental generation, or from other individuals belonging to the children's generation. Forms of transmission could also be one-to-one (as in many small-scale societies) or one-to-many (as in the education systems of complex societies).

Mathematical formulae could predict the rate of cultural change according to the mode of cultural transmission of a particular trait. For instance, cultural innovations are likely to spread much faster if the mode of transmission is one-to-many instead of one-to-one, or slower if it is vertical rather than horizontal or oblique. Similarly, when the predominant selective mechanism is frequency-dependent bias, we can predict that cultural innovations will spread rather slowly at the beginning, for few individuals will have adopted that particular cultural innovation initially. Then as it gradually becomes more popular, its expansion will accelerate. The predictions generated by the models could be measured against actual quantitative empirical data, so that particular instances of cultural change could be explained by the operation of one of those models or, more likely, their combined interaction.

In any case, the reasons that would explain why the transmission of cultural knowledge took place through those particular paths remained somewhat obscure. It could be argued, for instance, that behind model-based and frequency-dependent biases there was some evolutionary logic. I am more likely to adopt a successful cultural trait if I copy a prestigious individual than if I copy a less prestigious or low-status one. And the same applies to the frequency-dependent bias. Note that in both cases a 'successful cultural trait' is one that enhances the individual's biological fitness. Those with a hard-wired predisposition to imitate the majority, and/or successful individuals, will have better chances of passing on the genes that make them behave in that way. A similar argument, perhaps, could be put forward as regards content biases and guided variation. Maybe hard-wired into our cognitive system there is a predisposition to evaluate cultural practices and to decide whether to follow them or to change them according to one's biological interests. But beyond explorations of this hypothetical predisposition, early cultural-evolutionist models paid little attention to the characteristics of human cognition as a determining factor in the adoption of any one cultural trait.

The unspoken conclusion of those early theories of cultural evolution seemed to be that cultures could replicate irrespective of the characteristics of human cognition. But a closer look provides us with a slightly different picture. In fact, their implicit (and simplistic) model of human cognition viewed all cultures, almost by definition, as intuitive. Their underlying assumption was that cultures could be adaptive or maladaptive, but that they had to align with innate human predispositions – which, we must presume, had been selected for during the process of human evolution for their overall adaptive effects. And it was an (occasional?) incongruity between the adaptive value of those innate predispositions and their behavioural results that gave rise to maladaptive cultural practices. It might have been beneficial to my ancestors, that is, favourable to their reproductive fitness, to imitate successful individuals; therefore we all have a hard-wired predisposition to behave in this way. But sometimes that predisposition might misfire and we end up copying maladaptive practices. That is precisely what explains the discrepancy between biological evolution and cultural evolution and, from there, the need for a theory of cultural evolution.

Cognition hits back

The missing factor in this scheme is, as we have seen, a more sophisticated model of human cognition. By introducing that model to the explanation of cultural evolution, a more complex analysis becomes available. As we have seen, cultural traits can be sorted into adaptive and maladaptive, intuitive and counterintuitive. But notice that the four theoretical possibilities allowed in that matrix are not equally realistic in cultural-evolutionary terms. At one end, we could place those cultural practices that, *a priori*, seem to enjoy the best chances of survival and reproduction; namely, those that are simultaneously adaptive and intuitive. Our brain predisposes us to behave in a way, that is, to adopt certain cultural practices, that makes us have more children, so that predisposition and the cultural practices that it favours, is likely to proliferate without further ado. At the other end, by contrast (and *a priori* too), we would find the cultural practices with the worst chances of survival and reproduction: maladaptive and counterintuitive. Our brain predisposes us to reject those cultural practices that made us have fewer children. But in between those two extreme theoretical alternatives we have two interesting midway combinations. On the one hand, there is the maladaptive intuitive, which is what the models we have just seen were implicitly pointing at. Cultural evolution might diverge from biological evolution when we are misled by our innate predispositions to adopt a cultural behaviour that turns out to be maladaptive. And then we have the fourth possibility: adaptive counterintuitive cultures, which is when our innate predispositions discourage us from adopting an adaptive cultural practice.

In actual fact, this fourth possibility could not be contemplated in those early theories of cultural evolution precisely because of the unsophisticated model of human cognition they implicitly applied, a model in which the possibility of a counterintuitive culture was theoretically inconceivable. That might not be too bad when that counterintuitive culture happens to be maladaptive also. For it could be argued *a priori* that a maladaptive counterintuitive culture has few chances of being evolutionarily viable in any way – though this does not mean that it is utterly impossible (remember the compulsory celibacy of Catholic priests). But the adaptive counterintuitive, by contrast, poses an interesting paradox. For if that is evolutionarily possible at all – and, for the sake of the argument, let's assume that it is – we certainly need a slightly more sophisticated theory of cultural evolution to account for that possibility, otherwise we fall into a vicious circle of sorts: we need more culture to make counterintuitive culture cognitively acceptable. But how do we explain the existence of this surplus culture, then? In other words, what could induce human societies to produce the cultural means needed in order to adopt cultural practices that happen to be adaptive and counterintuitive at the same time? Do they need to be aware that those apparently unseemly cultural behaviours are good for them in the long run, as the doctor who advises her patient to go to the gym simply because 'she knows best'? Note that, in a way, this was the point of departure of the old functionalist approach to the study of culture in anthropology. Cultures exist because they fulfil some function. So no matter how irrational a particular cultural practice happens to be (namely, counterintuitive), if it exists it must be because it satisfies an important need (namely, adaptation). The problem with functionalism, as we saw, was the difficulty of defining those needs: were they social needs, individual needs, biological, psychological, psycho-biological? But thanks to evolutionary biology and psychology, we might be in a better position now. To restate our problem: how can (adaptive) counterintuitive cultural practices exist and successfully proliferate?

Cultural group selection

The cultural group selection hypothesis offers a possible answer to that question. In order to understand cultural group selection we need to define 'group selection', and understand why mainstream biological evolutionary theory rejected it, and ultimately why current cultural evolutionists have resurrected it. As we saw in Chapter 1, mainstream biological evolutionary theory is now gene-centred, which means that genes are defined as the unit of biological evolution. According to this point of view, adaptation is simply whatever happens to enhance the production of copies of a particular DNA sequence and, conversely, maladaptation is whatever happens to inhibit that same process. Now the interesting thing about this gene-centred view of evolution is that what is good for the replication of an individual's genes might not be good for the very same individual who possesses those genes: it might even kill him or her!

Altruistic cooperation as an evolutionary problem

The initial problem that the gene-centred view of evolution tried to address was the existence of animal behaviours that seemed to be at odds with the biological interests of the very same animal that performed that behaviour, specifically, cooperative behaviours in which animals seem to be helping each other instead of competing with each other. This type of behaviour is inconceivable in an individual-centred view of evolution, for if evolution only favours individuals that are good at surviving and reproducing themselves, any individual that helps other individuals instead of helping itself will have fewer possibilities of survival and reproduction than its more selfish counterparts and, therefore, will have fewer possibilities of passing on its altruistic genes. The problem is, however, that cooperation really does exist in nature. There are innumerable instances of organisms apparently helping each other. Take the example of a wolf pack hunting a big animal, when no wolf would have succeeded in bringing down that big prey on its own. Isn't that a credible instance of cooperation? Or look at ants in a colony, working tirelessly for the benefit of the whole community. And what about birds feeding their young, and sometimes going as far as risking their own lives to protect them against predators. This does not seem to agree with the sort of red-in-tooth-and-claw image of the natural world popularised by certain versions of Darwinian evolutionary theory.

Note that in these instances of cooperation, the individual members of what we might call the 'cooperative unit' thrive better by helping each other than by working in isolation. That is to say, they have better chances of survival and reproduction (and hence of passing on their cooperative genes to the next generation) by behaving cooperatively than by behaving selfishly. How can we explain the evolution of these cooperative behaviours from a Darwinian point of view? How could the principle of the survival of the fittest, which is supposed to be the corollary of natural selection, agree with that form of cooperation? Maybe what was wrong with the traditional individual-centred view of evolution was precisely the fact that 'the fittest' had to be an individual, that is, that the individual was the unit of selection in the evolutionary process. But what about if we posit the group as the unit of selection instead of the individual? Then individuals working in cooperation within a group would thrive better than those working in isolation simply because it is the group itself that turns out to be the fittest. Competition in nature is as merciless now as it was before, but competition occurs between groups, not between individuals. And the fittest groups are clearly those whose members cooperate and help each other. This is how cooperation in the natural world can finally be explained.

This is what group selection theory amounted to when it was initially formulated in the 1960s. But it did not take long for mainstream evolutionary theorists to realise that the group selection approach to evolution was hopelessly flawed. If we look more closely at cooperation in the natural world, which is what group selection was meant to account for, we will see that those very same cooperative or altruistic behaviours can actually be explained in a different, and better, way. To begin with, the instances of cooperative behaviour that we mentioned above – the wolf pack, the ant colony and bird parents feeding and protecting their chicks – are in point of fact rather different. So we might as well start by drawing some important distinctions as regards the nature of cooperation among non-human animals.

On the one hand, we find some form of cooperation in herd hunters such as wolves or lions. Wolves and lions live in herds and when they hunt big animals they do it collectively (particularly lionesses). So in a way it could be said that they 'cooperate' with each other. But that has nothing to do with altruistic cooperation. On the contrary, we may call it 'selfish' cooperation, for each individual animal pursues its own objective irrespective of what the other animals do. But the combination of each animal's individual selfish actions benefits them all, since they could not have hunted a big animal in isolation. Humans also have this type of cooperation. Think of a capitalist market: individual producers try to produce a product of the highest possible quality at the lowest possible cost. Why? Because this is how they maximise profit. But everyone benefits from the overall result: the production of cheaper and cheaper goods of higher and higher quality. We call this selfish cooperation because each individual cooperator pursues his or her own selfish interests. Only the overall result of the interaction of all those individual selfish behaviours turns out to be cooperative. So we do not need to posit the existence of cooperative genes or a group selection process that would make adaptive the unselfish behaviour presumably promoted by those genes, because cooperation in this case is merely the objective result of the interaction of subjectively selfish behaviours.

But besides this selfish cooperation (which many would not even call 'cooperation'), there is what we define as altruistic cooperation. Here cooperation is not based upon selfish acts but upon altruistic acts. These are acts detrimental to the actor's biological fitness (detrimental to the actor's capacity to survive and reproduce) in favour of someone else's. This is much less common in the animal kingdom, but there are still two clear instances where we can find this sort of altruistic cooperation among non-human animals.

One is reciprocal altruism. This is the situation where two individuals help each other in doing something that is profitable for each of them but that they could not do in isolation. The classic example is reciprocal grooming among chimpanzees. Chimpanzees spend long hours removing parasites form each other's bodies. This behaviour momentarily reduces the biological fitness of the one who performs it (I get no direct benefit while I am grooming another individual), with the expectation that shortly afterwards that other individual will return the favour. Note that this is different from the selfish cooperation we saw earlier because in this case we do actually find an animal helping another, even though it is clearly an interested form of help (maybe we should not even call it 'altruistic') for the helper will receive a benefit in due course. Reciprocal altruism among non-human animals turned out to be more problematic than was initially thought, for many instances of alleged reciprocal altruism fit better into the category of selfish cooperation. That is the case of inter-specific symbiotic behaviours that we find between small cleaning fish and the large fish they clean, for instance.

The second instance of altruistic cooperation that can be found among non-human animals is kin selection. This is much more widespread in the animal kingdom than reciprocal altruism, and this is clearly altruistic from the point of view of the individual. We find this kind of altruistic behaviour in mothers to their young, specifically in the case of mammals and birds. And also in the cooperation that we find among so-called eusocial insects: ants, bees and termites. It is called 'kin selection' because cooperation takes place among close kin, be these mothers and their young or all the individuals who live in the same colony in the case of insects, who happen to be brothers and sisters. The selective advantage of this form of cooperation was discovered by British biologist William Hamilton in the 1960s. That was precisely the time when the group selection hypothesis appeared on the scene, so kin selection can be seen as a final death blow to that approach.

How does kin selection actually work? Despite the mathematical sophistication of Hamilton's initial formulation, the rationale behind kin selection is in fact quite simple. Among animals who live very close to each other such as eusocial insects or those that are dependent on each other for their survival, such as bird and mammalian mothers and their young, individuals that help a closely related individual are more likely to spread their genes than those that do not, even though doing so might endanger their own survival as individuals! For instance, a mother that sacrifices herself for her young is more likely to spread her genes (including the gene that predisposes her to behave in this way) than one that does not, for the very simple reason that each of her offspring shares half of its genes with its mother. This means that her genes will be better off if she sacrifices herself for more than one of her children, rather than just one. Since she shares 50 percent of her genes with each of them, chances are that the gene or genes that promote this self-sacrificing behaviour will also be in the genome of at least one of those children. And the same applies to eusocial insects. Sisters who sacrifice themselves for each other and their queen spread their genes much better than those who don't, because the queen is in fact the only one that reproduces herself, and shares half of her genes with her sons and daughters but 75 per cent with her sisters, since they are haplodiploid (ants, bees and wasps); males are born out of unfertilised eggs and therefore have half as many chromosomes as females). Hamilton coined the phrase 'inclusive fitness' to refer specifically to biological fitness derived from kin selection. Whereas individual biological fitness amounts to an individual's capacity to survive and reproduce *itself*, inclusive biological fitness refers to its capacity to reproduce its genes, even at the cost of its own individual life.

Cooperation among humans is even 'worse'

We see therefore that none of the famous examples of cooperation or altruism require us to posit the hypothesis of group selection. Rather, a gene-centred view of evolution can explain them all. Behaviours that look altruistic from an individual point of view are perfectly selfish from the gene's point of view. From here came the title of Richard Dawkins's celebrated book *The Selfish Gene*, the gospel of the gene-centred view of evolution since its publication in 1976. As has been pointed out, most evolutionary biologists take this view, though some recalcitrant group selectionists still crop up from time to time.

But what about human beings? Is there anything special in humans' cooperative behaviours? Selfish cooperation, reciprocal altruism and kin selection all exist among humans as well. But humans also engage in a unique form of cooperation: cooperation between unrelated individuals that does not involve immediate reciprocity. In other words, a form of altruistic cooperation that is exactly the same as that of kin selection but without any kinship relationship linking the cooperators. How is this possible?

The absence of this form of cooperation among non-human animals made biologists think that it might be evolutionarily impossible. Take as an example the game called the prisoner's dilemma. Suppose you and a friend are caught by the police and accused of murder. One of you is the murderer but the police do not know which one. So they offer you the following deal: If you both keep silent, you'll get two years each for covering up the crime. If you blame each other, you'll get five years each, because this would indicate that one of you was the perpetrator. If one accuses the other and the accused keeps silent, the accuser is set free and the accused is jailed for 20 years, because the accused's silence will be interpreted as an admission of guilt. Theoretically, the best option for you both is to keep silent. But how can you be sure that your friend will keep silent too and not accuse you at the last minute? So what everybody ends up doing is accusing each other and getting five years each.

This game shows the impossibility or at least evolutionary instability of cooperation. There is the strong probability that those who would be willing to cooperate (in this case, keep silent) would end up paying a higher price, because there is no guarantee that the other person will do likewise when there is so much to be gained by doing precisely the opposite. So the most evolutionarily stable option is always to defect: to promise one thing and do otherwise. If all those who were so naïve as to keep their promises (keep silent) are wiped out by natural selection (end up in jail for 20 years), eventually only defectors survive. Paradoxically, groups of cooperators fare better than groups of non-cooperators, because cooperators will get away with only two years in jail whereas non-cooperators get five. The problem is that there is no way to enforce a cooperation agreement. Under these conditions, the only way the two prisoners could cooperate would be if a supervisor checked that both were honouring their agreement. In real life, that is precisely what has enabled human societies to develop this kind of cooperative behaviour: the existence of some sort of supervising entity (state, political power, etc.) that makes people behave in a cooperative way against their immediate selfish interest. That is why we all (or at least most of us) pay taxes. Even though we all know that taxes are needed to provide for collective goods (police, hospitals, roads, schools, etc.), very few people would pay their taxes if there were no authority with the power to punish tax cheats.

The apparent conclusion is that altruistic cooperation between non-kin is only possible among humans thanks to supervising entities capable of punishing defectors. But humans have lived for much of their evolutionary history in very simple societies, without overarching forms of authority powerful enough to enforce this sort of rule. This condition seems to imply that cooperation could not exist in simple ('pre-state') societies. However, what we know about hunter-gatherer groups (the predominant way of life in human EEA) does not bear out this hasty theoretical conclusion. Hunter-gatherers regularly practise sharing among unrelated individuals, though these have to be unrelated individuals belonging to the same band – a very important qualification, as we will see later. We can think of such sharing as an instance of the adaptive counterintuitive behaviours that we saw above. Cooperative groups do actually thrive better than uncooperative ones *in strictly biological terms*; they make better use of resources and therefore they have better chances of survival and reproduction. But there does not seem to be any possibility for a cooperative gene to evolve, because defectors will always have the upper hand in their interactions with naïve cooperators – that is what the prisoner's dilemma is supposed to have demonstrated. Therefore, if this form of cooperation exists in all human societies, and exclusively among humans, there must be something unique to humans beyond genes that can account for its occurrence.

The cultural factor

We already know that culture, humans' distinctive form of knowledge, can make individuals behave against their naturally selected instincts – the counterintuitive culture that we have been talking about. All that is needed is a special means to enforce cognitively unpalatable cultural messages. We asked why any society would bother doing that. Do their members have to 'be aware' that it is a good thing, like the doctor who knows what is good for our health and thus tells us to go to the gym, throw out our cigarettes and stick to a diet? Well, maybe not. So much the better if the members of that cooperative society explicitly place a strong moral value on cooperative attitudes, or perhaps they have made complex calculations which show that the most efficient way of maximising the resources they have at their disposal is by cooperating with each other instead of competing with each other. So much the better if all those cultural ideas are going through their minds while they are helping each other out. But in fact there is no need for it. There might be an objective process going on here, akin to the process of natural selection in biological evolution: non-human organisms do not have to be aware of the benefits of adaptation in order to produce adaptive traits and behaviours – many of them do not even have brains to be aware of anything at all. Mother Nature (i.e. natural selection) takes care of that. Natural selection may also be responsible for the enforcement of particular behaviours that happen to be adaptive but cannot be encoded, for whatever reason, into the individual's genome.

Note that this is different from the model of cultural evolution originally envisioned by Cavalli-Sforza, Robert Boyd and colleagues. Here we are talking about counterintuitive culture; altruistic cooperation with unrelated individuals, whereas the distinction between intuitive and counterintuitive culture was conspicuously absent in those early models of cultural evolution. This is an important qualification because accounting for the reproduction of intuitive cultural practices is simple (that was precisely what Cavalli-Sforza, Robert Boyd and colleagues were trying to accomplish, in particular for maladaptive intuitive practices); but it is the adaptive counterintuitive practices that remain unexplained. For if natural selection enforces upon us behaviours that go against whatever happens to be genetically encoded in our brains, natural selection is doing something different from what mainstream (gene-centred, Darwinian) models of biological evolution say it does: it makes us go against our genes! Isn't that what group-selectionist theorists had been proposing all along? Not quite. Group selection was a way of using Darwinian natural selection to explain behaviours that apparently did not go against the genes' interests but against the individual's biological interests. Group selectionists argued that this was possible because the group and not the individual (or his or her genes for that matter) is the unit of evolution. In any case, altruistic cooperation seems to fit perfectly well into this model (it benefits the group at the expense of the individual), hence the resurrection of the group-selection hypothesis in modern theories of cultural evolution. But there is a very important difference. Now the phenomenon in question is not simple, biological, group selection but rather *cultural* group selection.

Let us see how cultural group selection works and how it differs from biological group selection. Imagine a set of competing cultural groups with different cultural practices, some adaptive and some maladaptive. Suppose that the adaptive practices are adaptive for the group but maladaptive for the individual (i.e. altruistic cooperation). The cultural group that takes the collectively adaptive practice will outreproduce the others because that is what 'adaptive behaviour' actually means. But what about defectors? Remember that this was precisely the problem that the group selection hypothesis could not solve. A short-term individual sacrifice to the benefit of the group (by paying takes, for instance) is easy to conceptualise because in the long run we will all be better off. Until someone chooses not to pay and yet enjoys the benefits of the others' collective effort. From then onwards, the cooperative group might outreproduce uncooperative ones but, within the cooperative group, the uncooperative individual outreproduces his or her cooperative fellows. Therefore the cooperative group disappears not as a result of competition with non-cooperative groups but because of an internal conspiracy of uncooperative rebels.

But this problem of free-riders (uncooperative individuals within cooperative groups) does not apply to cultural group selection, because in cultural group selection all individuals have the same genetic predisposition for selfishness. Individuals cooperate within the group not because they are genetically predisposed to do so but because their culture makes them do it. Uncooperative individuals are in no way genetically different from the cooperative ones. So in this new context, the hypothetical fact that free-riding uncooperative individuals take unfair advantage of their cooperative neighbours is not a biological fact but a cultural fact. This distinction is very important. It is true that before the rise of states, the cooperative cultural group would appear to be as vulnerable to selfish rebels as the biological one. But something else is going on here now. All groups, cultural or biological, compete with each other for resources. But in the absence of culture, competition between biological groups – such as a wolf pack or a band of chimpanzees – can do very little to increase cooperation other than by changing their genes, for instance, by replacing selfish genes with altruistic cooperative genes. And we already know that a group of individuals genetically predisposed to cooperate is vulnerable to an internal conspiracy of selfish mutants. However, things are different when cooperative behaviours are enforced by cultural norms instead of genetic predispositions. In this case, as we have just seen, there is no danger of an internal conspiracy of selfish mutants because, in terms of their genetically inherited behavioural predispositions, all the group members are equally selfish to begin with. Because cooperative behaviours in this case are cultural behaviours, they do not depend on the existence of cooperative genes; we only need a way of preserving that unnatural (i.e. counterintuitive) cooperative culture, in the absence of a state that can punish transgressors. According to the cultural group selection hypothesis, this is what competition between cultural groups produces; the appropriate (cooperative) cultural rules are selected for and all the rest are selected against.

Otherwise stated, counterintuitive adaptive cultures survive and proliferate because there is a process of natural selection that picks up adaptive cultural traits (instead of genes). And that can only take place, needless to say, in a cultural species like the human species; no natural selection of cultural traits can take place when there are no cultural traits to be selected. Among the cultural traits that favour the inclusive fitness of individuals are those that promote cooperation within a group. Note, incidentally, that we are referring to altruism within a group – 'patriotic altruism' as it is sometimes termed – not generalised altruism with strangers. This latter form of altruism is a non-starter in evolutionary terms, no matter how morally laudable it might look to us. That is why sharing among hunter-gatherers, as we saw above, always takes place within the limits of a band. This is the flip side of cooperation in human evolution. It is very strong and remarkable but always within the limits of a group. Outside the group there is no cooperation but rather ruthless competition between internally cooperative groups.

The actual contents of that counterintuitive cooperative culture might vary. It could be fear of punishing gods, as some authors have contended: people behave altruistically towards their neighbours because they are afraid of supernatural punishment. In any case, because the behaviour enforced by that culture is counterintuitive, some special means will be needed for its assimilation, that is, rituals in the case of those punishing gods. But this is no longer a circular argument, for we already know why human societies would go to the trouble of enforcing those cognitively costly behaviours and beliefs upon their members. Those who did not or did not do it efficiently enough were wiped out by natural selection. That is what the cultural group selection hypothesis is all about.

Memes and cultural groups

The cultural group selection hypothesis enables us to explain the reproduction of individually maladaptive cultural practices that happen to be adaptive at the group level, which means that they are individually maladaptive in the short run but individually adaptive in the long run. The problem is that biological evolution is always short-sighted: whatever is good for my genes must be so right now, that is to say, it must enable me to have more children than other individuals with a different genotype or, at least, to have more grandchildren. But the reproductive advantage should not be effective in the too distant future; otherwise I – or my immediate descendants – might have been outreproduced well before the effects of that reproductive advantage can be felt. This is why the biological group selection hypothesis had to be discarded.

Those who actually outreproduce long-term adaptive mutations are what we have called 'free-riders' or 'defectors', who find themselves in the opposite situation: they behave maladaptively in the long run (at the group level) but adaptively in the short run (at the individual level). Now culture is precisely what inhibits the effects of the free-riders' behaviour by coercing selfish individuals into cooperative behaviour, and hence blocking the long-term deleterious consequences of their short-term advantages. But note that, in evolutionary terms, what gains selective benefit in cultural group selection is not the individual or the gene but the cultural group, which is by definition a hybrid entity made up of individuals, plus their genes and their cultural knowledge. So a cultural group entails a symbiotic alliance between different types of organisms in which the advantages enjoyed by one of the partners (human individuals plus their genes) are only long-term advantages, whereas the advantages enjoyed by the other partner (cultural knowledge) are short-term advantages: culture is reproduced with each generation. But what if those long-term advantages eventually disappear? In other words, what if that which began as a symbiotic relationship becomes a parasitic one?

The problem is that in this long-term perspective the line that separates the symbiotic from the parasitical gets very blurred. Maybe what explains the success of counterintuitive cultures is not the fact that those who adopt them have more children, or perhaps more grandchildren, or great-grandchildren (remember that this is a long-term effect) but merely that they end up reproducing more copies of those particular cultures in the next generation. Instead of passing on more genes to the next generation they pass on more 'memes'. The long-term reproductive advantage of those short-term maladaptive practices, insofar as it can be very long term, makes the difference between the symbiotic and the parasitic hardly visible. Obviously, to count as an example of long-term reproductive advantage, the cultural group has to survive and displace (outreproduce) the other cultural groups that did not adopt that particular cultural practice. But how do we know if what is actually being displaced is the cultural group and not its cultural practices?

A mind full of parasites

Take the example of religious rituals and beliefs, widely and deeply analysed in the anthropological and, more recently, cultural-evolutionist literature. At first sight, religious beliefs seem to be hopelessly counterintuitive: how could anyone believe in the existence of invisible beings, which are conceived as almighty or at any rate capable of the most bizarre deeds? And on top of that, how could anyone be persuaded to perform apparently useless actions, such as religious rituals, which are sometimes very costly and even self-damaging? The conventional explanation in the social sciences was to refer to some key social function that those strange ideas and behaviours were supposed to fulfil, such as the production of social consciousness or solidarity. We can easily reformulate this old functionalist explanation in cultural-evolutionist terms: the groups that happen to perform those rituals and uphold those beliefs outreproduced all the others. Consequently, the majority of societies have or have had some form of religious ritual and belief up to the present, because those who did not were simply selected against by the process of cultural evolution.

But could it not be the case that what was actually being reproduced were the religious cultural practices themselves rather than the cultural group that adopted them (i.e. those cultural practices plus human individuals and their genes)? If the reproductive advantage presumably enjoyed by individuals (and their genes) who had religious beliefs and practised religious rituals turns out to be a very long-term advantage, maybe that is because that hypothetical reproductive advantage does not really exist. What we have instead is a form of cultural knowledge, made up of religious behaviours and beliefs, which has successfully hijacked human brains to work for its own replication. To this effect, human individuals must be kept alive somehow and allowed to reproduce, but it is unclear the extent to which that survival and reproduction is to the actual benefit of those individuals or to the benefit of their cultural lodgers. Look at the biological world, with the innumerable parasites, commensals and mutualists that inhabit the body of another organism. It might not be very clear whether the relationship between host and guests is really mutualistic, which is when both host and guests benefit from that relationship (the cultural group selection hypothesis), commensal (only one benefits without harming the other), or parasitic (one benefits at the other's expense).

This is the perspective on cultural reproduction put forward by the so-called 'memetic' approach to the study of culture. We should treat cultures *as if* they were living organisms, mutualists, commensals or parasites, as the case may be, living inside the human mind. According to this approach, cultures really do evolve, pretty much like their living counterparts – the actual mutualists, commensals and parasites that dwell in another organism's body – but they do so for their own benefit, not ours, even though occasionally (or perhaps quite often) we also might reap some rewards out of that relationship. All this is very clear. But then there is an old question we should address head-on: what happens when those cultures turn out to be counterintuitive? What if our cultural guests find their brain hosts totally unsuitable for their reproduction and yet they manage to reproduce themselves in this unfriendly environment? We need special cultural means to assimilate counterintuitive culture. But we cannot account for the production of these cultural means without falling into a vicious circle. Remember that this was precisely the paradox that the cultural group selection hypothesis was meant to solve.

Well, perhaps in an already meme-infested brain certain counterintuitive messages become more cognitively palatable than otherwise would have been the case. Take the example of the teaching of maths. Only an expert minority can be very good at maths, but for a substantial portion of Western-educated populations, acquiring some mathematical knowledge is much easier than for those who have not had a Western education. It is as if the memes themselves could transform the human mind-brain so that what was initially cognitively costly eventually becomes much less so thanks, precisely, to those transformations.

Is this still a circular argument? Not if the transformations that the mind-brain goes through after successive meme infections enable the acquisition of new forms of (counterintuitive) cultural knowledge as a by-product of those memes, not an adaptation. Because it is a by-product, we do not need to look for an explanation as to why societies create a particular cultural environment congenial to the learning of counterintuitive cultural practices, for this merely takes place as an unintended consequence of that cultural environment. In other words, maybe the general mathematical knowledge assimilated by ordinary people in Western societies, counterintuitive as it may be, does not confer upon those people any selective advantage, but is merely a spin-off of an education culture that, on the whole, does confer that advantage. Thus, mathematical memes take advantage of minds already infected by that education culture in order to make copies of themselves. This does not mean that cultures can eventually disentangle themselves from the cognitive systems that enable their reproduction, that is, that cultural evolution goes its own way, independent from biological evolution. That could only happen if the reproduction of brains took place through Lamarckian, rather than Darwinian, inheritance. In that case, brains would be born already encultured and, consequently, cultural evolution could proceed alongside exclusively cultural paths because the cognitive system upon which it depended would be a culturally constructed cognitive system. But this is not how human brains (or any other biological organ for that matter) reproduce. Humans in any society have been born with the same brains for at least 200,000 years. So we are always confronted with the same challenge at each generational replacement: to transform that biologically constructed 'hunter-gatherer' brain into a culturally constructed brain suitable to the particular society in which that new generation is born.

The trouble with memes

The memetic approach has not gone unchallenged in the cultural-evolutionist literature. Two closely interrelated objections have been raised against this way of dealing with cultural reproduction and evolution. First, memetic theory entails a fax machine view of cultural reproduction: cultures go from mind to mind, from generation to generation, by making copies of themselves, as if the human brain was no more than a fax machine endlessly making copies of the same message once it has received the appropriate input. Genes do actually make copies of themselves in each new generation with only the odd error here and there. But it is clear that cultures do not reproduce themselves in this way. The fax machine model does not seem to correspond to the way in which human brains assimilate and transmit cultural messages. Human cultures are dynamic systems (if we can define them as 'systems' at all, another controversial issue we cannot address now), changing continuously as they go from generation to generation, and from mind to mind. Even though these are patterned, non-arbitrary changes, they make the analogy with genetic inheritance look rather unsatisfactory. The totality of the cultural knowledge of one particular generation in any society is never a photocopy of the cultural knowledge of the previous generation in the way in which the totality of that particular generation's genes is almost an exact copy, though recombined, of the totality of the previous generation's genes. This parallel between genetic and cultural inheritance put forward by memetic theory seems seriously flawed.

The second interrelated problem that the critics of memetics have brought up has to do with the very concept of meme. Some might argue that the cultural differences between two consecutive generations could still be suitably compared with their biological differences. In sexually reproducing species, new generations of individuals are never a 'photocopy' of their parents; genes and not individuals get copied in biological reproduction. In so far as those genes are recombined through sexual reproduction, each new generation of individuals is simultaneously different from and the same as the previous one: they are different as individuals though they have the same genes (making allowances for random mutations). Perhaps something similar happens in cultural reproduction: we inherit the same memes but we recombine them, sometimes at random, sometimes deliberately (an important distinction to which we shall return), plus the odd mutation here and there. But now we are confronted with the definition of the meme itself: what is a meme actually? To take the hypothesis of meme recombination seriously, we have to know what sort of thing we are recombining. Genes are DNA sequences that normally have or have had a function within the cell: the production of a protein, the switching on and off of other genes, etc. Genes can be measured and quantified; humans have around 30,000 genes. Genes can be seen through a microscope, we can even 'touch' them and alter them in precise ways. Nothing even remotely similar can be said about memes, which are ideas. How do we draw the limits of these ideas? How do we measure them? How do we quantify them? Until we find a proper answer for these questions, the hypothesis of meme recombination cannot make much progress. And hence the whole memetic theory of cultural reproduction does seem to rest on a rather shaky foundation.

But maybe these are not insurmountable problems. Darwin did not know how the process of biological inheritance actually worked when he formulated his theory of evolution though natural selection. The epidemiological model of cultural reproduction, of which memetic theory can be seen as an offshoot (perhaps a logical conclusion, perhaps a misleading corollary), continues to be a valid general approach to the study of cultural evolution. It may be an alternative to the cultural group selection model or just a complement to it. Cultures do create particular environments suitable to the reproduction of human individuals and to the reproduction of those cultures themselves, in whatever way we happen to parse them out. Perhaps altruistic cooperation exists among humans as a cultural behaviour that happens to be adaptive to its own survival, not ours. Maybe in the long run we also manage to reap some benefits from that. But what explains the short-term reproduction of that (initially) counterintuitive cultural behaviour is that it has colonised human brains already infested by other cultural lodgers, in such a way that the second wave of colonists (altruistic memes) can reproduce thanks to the transformations brought about in an initially hostile environment (the human brain made by selfish genes) by the first wave, the pioneer memes, so to speak. In other words, cooperative altruism becomes a by-product of an already encultured brain.

Culture-gene co-evolution

Cultural group selection means that groups with better (more adaptive) cultures outreproduce the others, so in the long run only those with adaptive cultures survive. According to memetic theory, by contrast, cultures evolve because they are adapted to their own survival, not ours. But in any case, in the meantime nothing happens, or nothing necessarily happens, at the gene level. Both cultural group selection and memetic theory are theories of cultural evolution. So what we explain with those theories is the evolution of particular cultural behaviours, specifically, of counterintuitive cultural behaviours, such as altruistic cooperation, which otherwise could not be properly accounted for. If nothing happens at the level of the genes, the net result of this process is the long-term reproduction of a cultural behaviour that is counterintuitive, no matter how adaptive (or not) it turns out to be in the long run.

Maladaptations are widespread in nature, but they are short-lived because evolution through natural selection eliminates them. In a way, adaptive counterintuitive cultures proliferate because of an initial biological maladaptation. We have to go on a diet to keep our cholesterol levels low because we eat too much saturated fat, and we do the latter because a craving for fatty foods, adaptive in our EEA where they were scarce, is no longer so in modern culturally constructed environments overstocked with that kind of food. We might speculate that perhaps in a not too distant future a person carrying a mutation that causes a dislike for fatty foods might have a better chance of survival and reproduction in this environment. Whatever the case, our diet culture might become a buffer of sorts against that genetic change. Diet cultures prevent saturated fat lovers from indulging their craving, thus the reproductive advantage of saturated fat haters is substantially reduced. (And let's set aside now how those counterintuitive diet cultures manage to reproduce.)

However, this has not always been the case in the cultural history of the human species. On several occasions particular cultural practices did give rise to a change in the genetic make-up of the individuals who adopted them. This is what is known as culture-gene co-evolution, or culture-led culture-gene co-evolution. Perhaps one of its most familiar examples is the evolution of adult lactose absorption in certain human populations. Lactose is a sugar present in most mammalian milks that needs special enzymes, called lactases, to be digested. All mammals become lactose intolerant when they reach adulthood, which means that none of them can digest milk after weaning. Lactose intolerance originates precisely in the disappearance of those enzymes from the small intestines of adults. There seems to be an adaptive logic to it: lactose intolerance facilitates weaning, enabling females to bear and feed more offspring. At the same time, it prevents adults from competing with their infants for food.

But all this is a relatively recent discovery that does not go further back than the 1960s. In fact, until then Westerners thought that milk was an essential food for all human populations, perhaps with the exception of a few odd individuals who did not seem to like it or tolerate it. At that time, when the rich West began to fathom ways of helping out famine-struck countries around the globe, milk and dairy products looked like the ideal means to combat food shortages everywhere because of their vitamins, proteins and fats. But much to the dismay of those wealthy benevolent Westerners, non-Western populations did not seem to like dairy products, even when faced with food shortages. Why was that? Cultural prejudice? Far from it. In fact, Westerners' fondness for milk and its derivatives turned out to be the anthropologically weird phenomenon; it is a clear instance of a culture-led culture-gene co-evolution.

Cultural practices that change our genes

The explanation was quite simple. In ancestral times, probably on the threshold of the Neolithic revolution, pastoralism became a viable mode of subsistence for certain populations, such as modern Northern Europeans' ancestors. Under those conditions, adult lactose intolerance became a maladaptive trait because it prevented adults from those populations from drinking their cattle's milk. Therefore, any mutant who did not lose his or her lactose-absorbing enzymes could thrive much better than his or her lactose-intolerant neighbours. Those mutants likely ended up outreproducing the others and, consequently, current Northern Europeans descend from those lactose-tolerant mutants. This is a genetic oddity, produced by the adoption of a certain cultural practice, such as pastoralism. It is true, on the other hand, that pastoralism has existed and still exists outside Northern Europe, most prominently in certain North African and Mediterranean populations such as the Greeks, among whom lactose tolerance is rather low. There is no need to go into too much detail here. Suffice to say that those southern populations had a cultural solution available, the processing of milk into low-lactose products such as cheese and yogurt, plus certain environmental conditions, such as strong ultra-violet B radiation, which together enabled them to absorb the required amount of calcium that their northern neighbours could only get from unprocessed milk.

Whatever the case, this example clearly shows that cultural evolution and biological evolution are not totally independent processes. In the same way as genes have created the appropriate conditions for culture to emerge and develop (we saw this in Chapter 1), cultures can also create particular environments, specifically long-term cultural practices such as modes of subsistence, wherein certain genes are more likely to reproduce than others. So we might be able to apply these insights to other biological peculiarities of the human species, such as the propensity for altruistic cooperation. Did they originate in a similar process of culture-gene co-evolution? That would explain very strange aspects of human behaviour observed by developmental psychologists, such as infants' propensity to help other people in rather simple tasks, that is, reaching some outlying object, when it is clear that they have not been taught to do that or they cannot have learned it in any way. If this is really an innate behavioural propensity of humans, how could it have evolved? The fact that this conduct can be seen in no other animal, not even our closest relatives, chimps and bonobos, clearly suggests that this is an evolutionary oddity. Now the particular way in which cultural knowledge intervened during the process of human evolution in order to produce this kind of behaviour, in apparent flagrant contradiction with the selfish-gene principles, is much less clear than what we have just seen for the lactose-tolerance adaptation.

It could have been a combination of culture-gene co-evolution plus cultural group selection. Perhaps altruistic cultural groups became so successful that they ended up transforming the genetic make-up of their members in such a way that they were not vulnerable to the attacks of selfish mutants. Alternatively, we might understand culture-gene co-evolution as gene-meme co-evolution. By becoming lactose tolerant it is the pastoralist memes that get better reproduced, not us. So memes not only hijack our brain to make copies of themselves but they can even change our genome! Note that this is not an instance of the dreadful Lamarckian inheritance that we mentioned above (dreadful as far as genes are concerned, not memes): it is not just a matter of passing culturally learned behaviours onto our genes. What happens is a little bit more complicated even though the effects might look (somewhat disturbingly) similar. Cultures do create particular environments that not only favour the reproduction of particular memes at the expense of others, but also the reproduction of particular genes. So apparently everything looks as if cultures did actually create those particular genes themselves. But that is not true: they only create the favourable conditions for their reproduction, which continues to be a Darwinian process through and through. In any case, if cooperative altruism in humans has evolved through a process of gene-meme co-evolution, that means that it is no longer counterintuitive, or that it is much less counterintuitive than it was at the beginning of human evolution and much less counterintuitive than it is for non-human animals now. This is so because the memes for cooperative altruism, by changing the environment wherein the genes that code for the particular characteristics of the human brain make copies of themselves, may have enabled the reproduction of an innately cooperative brain in humans.

Random processes and purposeful action

All modern theories of cultural evolution – adaptationism, cultural group selection, memetics and culture-gene co-evolution – seem to underscore the usefulness of a Darwinian approach in the study of culture. And even perhaps in the not too distant future they seem to point at a possible synthesis or consilience between the natural and the social sciences in the study of human behaviour. All this is very well but there is one last question in the theory of cultural evolution that might cast a shadow on this optimistic prospect, and that is the function of purposeful action in cultural evolution. We saw in the last chapter that biological evolution is the result of two processes: one contingent, the random mutations of DNA sequences, the other necessary, the natural selection of those mutations that happen to be adaptive. The possibility of using Darwinian theory to analyse cultural evolution has been criticised precisely in this respect. It is often argued that the mutations that take place in cultural artefacts are far from random, they are purposeful. Human beings create their cultures, and without a doubt they do not do it at random, they have a purpose in doing what they do. Hence the process of natural selection in biological evolution seems to be very different indeed from that which takes place in cultural evolution – random mutations in the one case, purposeful creations in the other. Does that mean that cultural evolution is less Darwinian than biological evolution, or perhaps non-Darwinian at all?

American philosopher Daniel Dennett has dealt with this question, particularly in his latest publication (see bibliographical note). Dennett's position is interesting because despite being an uncompromising Darwinian who has unambiguously embraced the memetic approach to cultural evolution, he openly recognises that cultural evolution becomes gradually less Darwinian. Cultural evolution starts off as a fully Darwinian process, with memes competing with each other to colonise human minds and thus produce copies of themselves, in exactly the same way as genes compete with each other for the same purpose. This is what Dennett calls a bottom-up process of 'competence without comprehension'. Successful genes will end up producing very reliable organs in the same way as successful memes will produce outstanding cultural artefacts. But neither genes nor memes need to have any understanding or comprehension of what they are doing. The clearest example of a marvellous cultural artefact that can be seen as the result of this bottom-up process is human language. Nobody has deliberately created any natural language; all natural human languages are the result of a long process of competition between different sounds to colonise human minds, or mind-brains and thus be able to reproduce. This is a perfectly Darwinian process of random mutation, natural selection and survival of the fittest, practically identical to the processes that take place in biological evolution.

But Dennett points out that it is precisely the production of this magnificent cultural gadget that enables humans to free themselves from the shackles of Darwinian evolution, for it is thanks to language that humans can begin to have competence *with* comprehension. From then onwards, alongside the blind bottom-up process of Darwinian cultural evolution we will have top-down process of intelligent design. He puts as an example of a cultural artefact that is clearly the product of intelligent design the famous temple of La Sagrada Família in Barcelona. We can compare Antoni Gaudí's renowned work with a termite castle, for both structures may be deceptively similar but their processes of production are radically different. Whereas the termite castle is the result of Darwinian biological evolution, none of the termites has any idea, objective or purpose in building their castle, La Sagrada Família is the result of Gaudí's intelligent design. Maybe the unskilled bricklayers who are actually building the walls and columns of Gaudí's church do not know much about the master's plans. But still, if we consider the whole set of activities that gave rise to La Sagrada Família, from Gaudí's initial blueprints to the motions of the humblest builder, it all looks very different form the activities of termites and, by implication, from the random mutations of a DNA sequence. Clearly, what applies to La Sagrada Família equally applies to any other contrived product of human culture. There is a purpose behind those products and a process of production, more or less meticulous and painstaking, as the case may be, that sets them quite apart from the random mutations that characterise biological evolution.

Should we conclude with Dennett, therefore, that cultural evolution becomes less Darwinian as time goes by, more akin to an intelligently designed process than to that of evolution by natural selection? Let me go back to another distinction with which I finished Chapter 3: culture from within and culture from without. I have argued that structuralism gives us a perspective on culture from within, as a meaning-producing mechanism. Human cultural behaviour can be explained in terms of its meaning, which in turn originates in the capacity of the performer to have done otherwise. The fact that I don't have sex with my sister is a culturally meaningful action in so far as my behaviour can be seen as the result of the incest taboo, a cultural norm, prevailing in my society, and, therefore, in so far as I could have done otherwise: I could have broken that taboo and slept with my sister. But if I don't have sex with my sister because I don't feel any sexual attraction for her, and that is the result of some innate incest avoidance, then that behaviour is in no way culturally meaningful. It is merely an instinctual behaviour and, as such, I could not have done otherwise; quite literally, 'it is in my nature' not to have sex with my sister but not 'in my culture'. Notice that when we look at culture from within, the actual behaviour that derives from a cultural norm is always indeterminate. Of course there is a high probability that if a particular cultural norm prevails in a society the majority of people will comply with that norm. But there is always the possibility that someone might disobey, and this is a crucial point, for if no one could actually disobey the principle that rules behaviour in that society would not be a cultural norm.

But what happens when we look at culture from without? Old functionalists thought that cultures exist because they fulfil a particular function. Modern adaptationists add that that function consists in adaptation to a particular environment. We know, however, that not all cultural productions are functional or adaptive. An explanation in terms of selective advantage for the individual or the group that adopts that particular culture seems to be appropriate for adaptive cultural behaviour, while an explanation in terms of the selective advantage for the cultural item itself would be adequate for maladaptive cultural behaviour. Still, others might argue that it is always the selective advantage for the cultural form that accounts for its replicability and hence for the actual production of whatever behaviour that cultural form is supposed to bring into being. But be that as it may, in none of these explanations does indeterminacy seem to play any role. In other words, the fact that people behave in a certain way because, in one way or another (probably unconsciously), they entertain the possibility of doing otherwise and finally they reach the conclusion that they might as well follow the cultural convention or norm of their society, the fact that it is this thought process that makes people behave the way they do is irrelevant to the explanation of culture from without.

If the thought processes that give rise to a particular behaviour are irrelevant when we look at culture from without, that means that it could have been a totally random behaviour and the result would have been the same. Adaptive behaviours for the individuals themselves or for the cultures that make them behave in that way will be selected in whereas maladaptive behaviours will be selected against. Whatever causes those behaviours 'from within' would not make any difference. Remember that, quite obviously perhaps, selection always acts upon behaviour and never upon ideas or thought processes.

Imagine that a clever scientist discovers that what we see now as random mutations in DNA sequences are in fact not random at all, that those mutations are the result of some conscious decision. Perhaps the genes themselves happen to be conscious agents that decide in what way they are going to mutate, taking into account that their objective is always to produce as many copies of themselves as possible. Needless to say, this is a totally fictitious scenario, there is no way genes can have any form of consciousness and, *a fortiori*, perform any deliberate action. But let me pursue this line of reasoning just for the sake of the argument. Now despite being conscious beings it turns out that genes are not omniscient. They are never a hundred per cent certain that a particular mutation is going to be successful. Some of them are quite clever and manage to pass on their mutations to the following generation. And among them there is a tiny minority, they call them geniuses, who have been able to replicate their mutations for a very long time, generation after generation. Others, by contrast, perhaps the majority, never succeed in such an endeavour, and most of their mutations are lost forever.

Is this a random process? If we look at it from within, from the point of view of the genes themselves, this is certainly not a random process. Each mutation is the result of a conscientious deliberation and hard work. Some genes work harder than others, no doubt about that. But none of them can be said to produce his or her decisions (let us suppose these genes have a gender) as a result of a random process. But what about if we look at it from without? Is this fanciful picture that I have just imagined, unrealistic as it may seem, totally different from the way in which cultural production takes place in human societies? Think of the hundreds of manuscripts that will never see the light of the day because the publishers have turned them down, or perhaps the hundreds of draft papers that will never be published because their authors never bothered sending them to a journal. And among those that manage to get published, how many of them will be read and will have a real impact on their audience, whatever that audience happens to be? And the same applies, it seems to me, to any other cultural product, be this a work of art, a scientific creation or a business venture. They are all the product of hard work and conscious decision-making, and yet in none of them does this seem to be a sufficient condition for their success, a very similar situation to the one in which our imaginary industrious genes found themselves.

Still some might be tempted to argue that the change in a DNA basis, which is how all genetic mutations start off, can in no way be compared with the deliberate production of a cultural artefact. In the one case, we have an almost instantaneous and microscopic event that does not seem to obey any rule other than that of the toss of a coin, in the other, a long and painstaking process of production. None of them has any guarantee of success, certainly, but still they definitely look very different indeed. However, this is only, at best, a partial objection, an objection that originates in a short-term view of biological evolution. Without a doubt, a successful genetic mutation is, in all probability, the result of hundreds of microscopic changes in DNA sequences that take place over several reproductive cycles. But if we look at genetic mutations from a long-term perspective, namely as a completed process that has given rise to a successful adaptation, the difference between that successful adaptation and the most sophisticated cultural product practically disappears, as British biologist Richard Dawkins perceptively observed in his celebrated comparison between a watch and an eye in *The Blind Watchmaker*. Of course, I am not trying to suggest that there is some sort of hidden intelligence that guides biological evolution unbeknown to us. In fact, it is the very opposite that I am hinting at: the intelligent design that seems to rule cultural evolution from within runs parallel to a random process when we look at it from without.

Bibliographical note on Chapter 4

As mentioned in the main text, classic works of cultural-evolutionary theory are those by Cavalli-Sforza and Feldman (1981) and Boyd and Richerson (1985). More fitting for those who have an aversion to mathematical formulae is Richerson and Boyd (2005). Both Mesoudi (2011) and Lewens (2015) provide an excellent overview of the research, which is comprehensive, interdisciplinary and apt for beginners, and which includes balanced theoretical discussions and examples from archaeology, ethnography and experimental psychology (see also Mesoudi et al. 2006 and Richerson and Christiansen 2013). Group selection theory in biology was initially put forward by Wynne-Edwards (1962), criticised by Maynard-Smith (1964) and Dawkins (1976), and supported by E.O. Wilson (1975) and Wilson and Sober (1994). Whereas the works of Wynne-Edwards and Maynard-Smith cater for more scientifically-trained readers, Dawkins and Wilson aim at a wider audience. Hamilton (1964) proposed the first formalisation of kin selection theory (again, a text unsuitable for beginners), developed further by Dawkins (1976) and Maynard-Smith (1964). See Holland (2012) for an excellent overview of Hamilton's thesis in the light of ethnographic evidence and for an insightful comparison between evolutionary biology and traditional anthropological theories of kinship. The memetic approach to the analysis of culture has not caught on in mainstream social sciences (so far), even though it is quite popular outside academic circles. The concept of 'meme' was originally formulated by Dawkins (1976), and further developments in memetic theory of cultural reproduction can be found in Aunger (2000), Blackmore (1999) and Dennett (2017). None of these works contains inscrutable jargon or off-putting mathematical formulae, but while Dennett's work is the most philosophically sophisticated, Blackmore's text constitutes perhaps the best general introduction. Critical views on the memetic approach can be seen in Atran (2001) and Sperber (2000). The relationship between cultural group selection and cooperation was put forward by Henrich (2004). Further developments and discussions can be found in Boyd and Richerson (2006) and Richerson et al. (2014). Both works include a target article and open peer commentaries by leading specialists. A classic work on culture-gene co-evolution is Durham (1991), which contains several instances of interaction between cultural and biological evolution, such as the case of lactose tolerance among Northern Europeans that we have just seen. Richerson and Boyd (2010) provide an up-to-date synthesis on this same issue. See also Chudek and Henrich (2011) for an application of the culture-gene co-evolution approach to the analysis of human cooperation.

References

Atran, S. 2001. 'The Trouble with Memes: Inference versus Imitation in Cultural Creation'. *Human Nature* 12(4): 351–381.
Aunger, R., ed. 2000. *Darwinizing Culture*. Oxford: Oxford University Press.
Blackmore, S. 1999. *The Meme Machine*. Oxford: Oxford University Press.

Boyd, P. and P.J. Richerson. 1985. *Culture and the Evolutionary Process*. Chicago: Chicago University Press.
———. 2006. 'Solving the Puzzle of Human Cooperation'. In *Evolution and Culture*. S.C. Levinson and P. Jaisson, eds. Cambridge, MA: MIT Press.
Cavalli-Sforza, L.L. and M.W. Feldman. 1981. *Cultural Transmission and Evolution*. Princeton: Princeton University Press.
Chudek, M. and J. Henrich. 2011. 'Culture-Gene Coevolution, Norm-Psychology, and the Emergence of Human Prosociality'. *Trends in Cognitive Sciences* 15: 218–226.
Dawkins, R. 1976. *The Selfish Gene*. Oxford: Oxford University Press.
Dennett, D. 2017. *From Bacteria to Bach and Back*. New York: W.W. Norton.
Durham, W.H. 1991. *Coevolution. Genes, Culture, and Human Diversity*. Stanford, CA: Stanford University Press.
Hamilton, W.D. 1964. 'The Genetical Evolution of Social Behavior I and II'. *Journal of Theoretical Biology* 7: 1–52.
Henrich, J. 2004. 'Cultural Group Selection, Coevolutionary Processes and Large-Scale Cooperation'. *Journal of Economic Behavior and Organization* 53: 3–35.
Holland, M. 2012. *Social Bonding and Nurture Kinship*. London: Senate House Library.
Lewens, T. 2015. *Cultural Evolution. Conceptual Challenges*. Oxford: Oxford University Press.
Maynard-Smith, J. 1964. 'Group Selection and Kin Selection'. *Nature* 201: 1145–1147.
Mesoudi, A. 2011. *Cultural Evolution*. Chicago: University of Chicago Press.
———, A. Whiten and K.N. Laland. 2006. 'Towards a Unified Science of Cultural Evolution'. *Behavioral and Brain Sciences* 29: 329–383.
Richerson, P.J. and R. Boyd. 2005. *Not by Genes Alone. How Culture Transformed Human Evolution*. Chicago: University of Chicago Press.
———. 2010. 'The Darwinian Theory of Human Cultural Evolution and Gene-Culture Coevolution'. In *Evolution since Darwin. The First 150 Years*. M. A. Bell et al., eds. Sunderland, MA: Sinauer.
Richerson, P.J. and M.H. Christiansen. 2013. *Cultural Evolution. Society, Technology, Language, and Religion*. Cambridge, MA: MIT Press.
Richerson, P.J., R. Baldini, A.V. Bell, K. Demps, K. Frost, V. Hillis, S. Mathew, E.K. Newton, N. Naar, L. Newson, C. Ross, P.E. Smaldino, T.M. Waring and M. Zefferman. 2014. 'Cultural Group Selection Plays an Essential Role in Explaining Human Cooperation: A Sketch of the Evidence'. *Behavioral and Brain Sciences*. Available on CJO2014. doi:10.1017/S0140525X1400106X.
Sperber, D. 2000. 'An Objection to the Memetic Approach to Culture'. In *Darwinizing Culture*. R. Aunger, ed. Oxford: Oxford University Press.
Wilson, D.S. and E. Sober. 1994. 'Reintroducing Group Selection to the Human Behavioral Sciences'. *Behavioral and Brain Sciences* 17(4): 585–654.
Wilson, E.O. 1975. *Sociobiology: The New Synthesis*. Cambridge, MA: Harvard University Press.
Wynne-Edwards, V.C. 1962. *Animal Dispersion in Relation to Behaviour*. Edinburgh: Oliver and Boyd.

5
SUMMARY AND CONCLUSIONS

1. Our first problem was to account for human diversity, the most distinctive characteristic of our species. Why is it that we humans are so different from each other? Human differences cannot simply be mapped onto biological differences (say, intra-specific genetic differences), since these latter are not bigger than those of any other species. So there has to be something else that accounts for those differences. And this something else is what we call 'culture', or 'cultural knowledge'. We are the only species capable of producing this very special form of knowledge in a massive, systematic and cumulative way. We have seen the historical-evolutionary reasons that may account for the emergence of culture in human evolution. These are biological reasons, since we humans are biological beings before cultural beings. So our biology cannot tell us about our differences, directly as it were, but it can tell us a lot indirectly. It cannot tell us what those differences are but it can tell us what causes those differences.
2. The rise of cumulative culture in human evolution was caused by a very distinctive capability that only humans have fully developed: mentalisation or Theory of Mind. Notice that 'having a mind' is not what distinguishes humans. If we consider that a mind is simply the repository of our thoughts (intentions, desires, etc.), most animals have minds as well – perhaps all living organisms do, one way or another. What makes us different is our capacity to imagine minds in other beings. As soon as we reach this threshold, the conditions for the rise of cumulative culture are laid out before us. If we can enter other people's minds we can appropriate their contents, which might have been appropriated from other people's minds, etc. In the mind of any human being, in so far as it is transparent to others, there will always be a collective mind, the mind of numerous other human beings. A similar thing happens with our genome. In the genome of any living organism we can see the genome of many other organisms as well.

3. The definition of culture as a form of knowledge has enabled us to make it commensurable with the forms of knowledge that humans share with other species. These are genetic knowledge and individual knowledge. Knowledge is any form of information that is processed by an organism and that produces some change in that organism; for instance, behavioural changes in the case of animals. Genetic knowledge is knowledge from the environment incorporated into our genes by means of natural selection, going all the way back in time through the evolution of our species and the species that preceded it. It can be very precise, effective and metabolically cheap. And it has only one 'purpose': to make copies of itself. The problem is that it is very rigid, especially for those organisms with long reproductive cycles (such as large animals). That is why those large organisms make use of another form of knowledge, individual knowledge. Individual knowledge is flexible but it cannot be accumulated and it is more expensive in metabolic terms: we need a special organ to produce it. Even more expensive is cultural knowledge, but it has a very remarkable advantage: unlike individual knowledge, it can be accumulated, so we might as well say that it clearly compensates for the extra metabolic expenses (big brains) that it requires.
4. From a mere evolutionary point of view, cultural knowledge has been very useful to the human species. It has enabled us to adapt – that is, live and reproduce – in practically all environments on earth without having to wait for a genetic mutation or having to process massive amounts of individual knowledge. But the concomitant effect of this exceptional form of knowledge is that it makes humans very different from each other. Since cultural knowledge is transmitted through communication between minds and we cannot communicate with the minds of all humans, cultural knowledge divides up the human species in groups with very loose time-space boundaries. These cultural groups, what we normally call 'cultures', are the origin of human diversity.
5. What makes cultures unintelligible to each other is similar to what makes two human beings unintelligible to each other. Why do we sometimes fail to understand another human being? Simply because we don't know what is going on in his or her mind. To the extent that our minds are 'collective' minds, our misunderstanding may have occurred because we don't know some of the 'other minds' that happen to be inside the mind of the individual we want to understand. This is what we might define as a 'cultural misunderstanding'. The problem is how you make yourself familiar with those other minds without having to go through all the experiences that enabled that particular person to have access to them, in other words, without having to live that other person's life – which is what being brought up in a particular culture amounts to.
6. Making yourself familiar with an alien culture is just the first step towards explaining cultural differences. How do we account for those differences? Different ways of answering this question have given rise to different anthropological theories. Maybe humans do not actually have different cultures but rather different amounts of one same culture. Just as a child's knowledge of

the world is different from that of an adult simply because the child has less knowledge than the adult. Or maybe humans do have different cultures after all, and it is all just a matter of different traditions. Because culture originates in human interactions, we have different cultures because we have different interactions. I have the culture that I have because I got it from my elders, that is what traditions are all about. But perhaps there is more to it than that; maybe different cultures help humans solve different problems.

7. We have considered four main anthropological theories: social evolutionism, historicism, functionalism and structuralism. Culture theories do not explain why humans have cultures but why humans have the particular cultures that they have. These four theories, or families of theories, try to account for human diversity by means of the concept of culture. But here the similarities stop. Social evolutionism was the oldest properly anthropological theory of human diversity. According to this approach, humans are different because of the different amounts of cultural knowledge they have accumulated throughout history. So-called 'primitive' societies are different from us because they have less history and therefore less culture. Modern societies are different from all the others because they have accumulated more cultural knowledge, due to their longer history. Historicists, functionalists and structuralists, the classic paradigms of modern anthropology, took a radically different path in explaining human diversity. Modern anthropologists consider all human societies to have the same 'amount' of history. Cultural differences between humans are not quantitative but qualitative. Historicists think that different histories have given rise to different cultures. Both social evolutionists and historicists see culture as essentially a historical product that requires a historical explanation. They differ in the sort of historical explanation they provide. Functionalists, by contrast, avoid historical explanations of culture. Humans have cultures because they are useful to them, not simply because they inherited them from their ancestors. The problem is how we account for that utility.

8. Structuralism was a first attempt, within classic anthropological theories, at superseding the opposition contingency/necessity, or historicity/usefulness in the analysis of culture. From this perspective, cultures are historical products and they are also useful to humans. But this is not how we can explain cultural differences. Cultures are systems of meaning and that is what makes them so different from each other. If we understand cultural meaning as a form of collective intention, we can see that those meanings, like the individual intentions of which they are made, originate in our capacity to do otherwise. Therefore we can discern the process of meaning production of any cultural practice by comparing it with its opposite, with what could have been done in its stead, either as a logical possibility or as a sociological reality that has taken place elsewhere or at another time. By emphasising the process of meaning production as an explanation of culture, structuralism achieved two aims. First, it made all human cultures commensurable with each other without turning them into more or less developed versions of each other. Secondly, in the structuralist

imaginary human cultures became sets of possibilities for human action logically arranged in systems of binary oppositions. This approach is equidistant from the historicist view that sees humans as blindly replicating whatever they inherited from their elders and the functionalist view that espouses a rationalist utilitarian view of culture.

9. We have already seen that cultures can be very useful to human beings in plain evolutionary terms. Therefore, one way of understanding this diversity of cultures is by looking at the ways in which each particular culture fulfils its function of enabling humans to live and reproduce in particular environments. This is a way of upgrading what we have called the functionalist explanation of culture by providing an objective concept of utility. But since cultural knowledge can reproduce itself irrespective of the reproduction of human beings, not all cultural knowledge necessarily has to be useful to human beings. Hence the possibility of defining cultures metaphorically as independent living organisms that sometimes enter into symbiotic relationships with humans, and at other times become parasitic on these very same human beings. Certainly, this is just a metaphor, since a culture is only a system of mental representations and does not exist in material terms. But the metaphor can be very telling. While we try to understand any particular culture we should take its double-sided nature into consideration. Functionalism throws into relief the symbiotic human-culture relationship: cultures exist because they are useful to humans. The parasitic relationship, by contrast, can be seen as a modern upgrading of the old historicist approach: it is humans that are useful to cultures by passing them down from one individual to another and from one generation to another.

10. The problem with these traditional ways of looking at culture is that they cannot account for cultural change. If cultures are just traditions that we learn from our elders and teach to our descendants, we cannot explain why those traditions happen to change sometimes. Maybe we just make mistakes in this process of learning-teaching. But then cultural history would be a totally contingent process, which clearly it is not. Something else must be going on there. Functionalists fared even worse in that respect, for their point of departure was that cultures were the way they were simply because they were useful to humans. But they could not explain how they got that way, nor could they explain why they changed despite being so useful. We need a theory of cultural evolution parallel to the theory of biological evolution. At first sight, this looks like an interesting comparison because, clearly, a very similar process of natural selection and adaptation seems to be going on in cultural history. This process explains why humans have been so successful in adapting to very different natural environments without changing their genes – and/or without massive investments in individual knowledge to produce the appropriate behaviours. But that is only half (perhaps even less than half) the story. Cultures are adaptive for humans, but they are also maladaptive. How do we explain the reproduction of maladaptive cultures? Natural selection cannot help us here.

11. A possible answer is to rethink the very concept of adaptation in humans. Perhaps some cultures are maladaptive in the short term but adaptive in the long term or, stated otherwise, they are maladaptive from the individual point of view but adaptive from the point of view of the society where those individuals live. That is the solution put forward by the cultural group selection hypothesis. According to this perspective, natural selection would act upon societies (i.e. cultural groups); hence what is maladaptive for the individual proliferates because it is good for the society. Furthermore, it could also be the case that those individual/short-term biological maladaptations are only temporary. Societies and their cultures create particular environments that differ from the natural environments in which humans' current biological characteristics evolved. So natural selection can also operate in these new culturally constructed environments by picking up new biological characteristics for humans; new characteristics that would make them less maladapted in this new context. Therefore cultural group selection might give rise to culture-gene co-evolution.

12. The other possible answer to the question of why maladaptive cultures proliferate also entails rethinking the concept of adaptation. But this time it is because we apply it to the cultures themselves, not to humans. Maladaptive cultures proliferate because they are adapted to their own survival and reproduction, not ours. Now we need to know what 'survival and reproduction' for a culture actually means. It means being able to colonise human brains, or human mind-brains. For cultures 'live' in human minds and 'reproduce' by getting into other human minds. Here is where the analogy between cultures and living organisms gains some purchase, specifically, with those living organisms that live at the expense of, or in conjunction with, another living organism (namely parasites, commensals and symbionts). From this perspective, we have managed to explain cultural change because cultures certainly evolve, the same as any other living organism. In this view, cultural evolution is the process that turns any cultural message – perhaps contingently created in the pandemonium of human history – into an appropriate guest for human minds. This is what the memetic theory of cultural reproduction proposes.

13. Many things remain unanswered in this account of the cultured nature of human beings. Most prominently among them is cultural transmission. We know how genes go from body to body through the process of biological reproduction. But we do not really know how cultural ideas go from mind to mind. Once the so-called fax-machine model of cultural reproduction has been discarded, the only thing we know for sure of is that cultural ideas are transformed as they go from one mind to another and they do it in a non-random way. The centrality of the notion of meaning in cultural analysis comes back into play here. Ideas can colonise human minds because those minds find them meaningful. But it is still this meaning-producing process that remains elusive. French anthropologist Dan Sperber (a student of Lévi-Strauss) has suggested that in cultural transmission, ideas are not merely replicated or

photocopied but they are actually *re*-produced, that is, produced anew following the tracks of what he calls 'cultural attractors', by which he means both the psychological and ecological factors that determine the cognitive feasibility of those ideas. This is a fundamental question, for just as Darwinism could only fully account for biodiversity once the mechanism of genetic inheritance was discovered, our explanation of cultural diversity will remain incomplete until we can identify a parallel mechanism of cultural inheritance. But that is surely a topic for another book.

INDEX

Page numbers in *italic* refer to figures.

adaptation 12, 13, 15, 16, 18, 42, 46, 49, 71, 82, 83, 91, 95, 100, 101, 106, 110, 114, 117, 118, 124, 125; behavioural 28, 29; *see also* maladaptation
adaptationism 74, 115, 117
altruism 101–108, 112, 114, 115; reciprocal 103–104; *see also* cooperation
altruistic cooperation *see* altruism
attachment theory 33
Australopithecines *see* Australopithecus
Australopithecus 9, *21*, 23–24, 27–28
autism 31–32, 34

belief 26, 32–33, 38, 46, 51, 54, 58, 75–76, 79–82, 86–87, 95, 108; in magic 79–82; religious 109; in spirits 45, 56
binary oppositions 87, 89–90, 124
biodiversity 71, 82, 126
biological anthropology 3, 7
biological evolutionism *see* evolutionism
biology 4, 8, 34, 46, 119, 121; evolutionary 4–5, 35, 97, 100, 119; human 3–4, 8, 62, 94
bipedalism 24, 27
Boas, Franz 73–79, 82–83, 86, 91–92
Boyd, Robert 62, 97–98, 106, 119
brain 5, 6, 8–9, 18, 20, 22, *23*, 25–31, 33–34, 39, 44, 46, 49, 51–56, 83, 85, 100, 106, 109–112, 114–116, 122, 125
bridewealth 89–90

Cavalli-Sforza, Luigi 97–98, 106, 119
celibacy 47, 100
chimpanzees 3, 7, 9, 20, *21*, 22–25, 29, 32–33, 103, 107
Chomsky, Noam 30, 49
chromosomes 12, 14, 104
cloning 11–13
cognition 26, 34, 76, 99–100
cognitive modules 27–31, 34, 49
cognitive processing 40–41
commensalistic 43, 110, 125
computer 14, 27, 29, 41, 56, 87
conscious 16, 53, 63, 65, 90, 109, 117–118; *see also* unconscious
consciousness *see* conscious
contingency 2, 5, 70, 75, 79, 82–83, 86–87, 91, 94–96, 115, 123–125; *see also* random
contraceptive sex 43, 47
cooperation 33–34, 39, 49, 63, 101–108, 112, 114–115, 119; selfish 102–104
counterintuitive knowledge 46–48, 95, 99–100, 105–110, 112–113, 115
creationism 10, 16; *see also* intelligent design
cultural evolution *see* evolution
cultural groups 48, 68, 101, 106–110, 112, 114–115, 119, 122, 125
cultural history 5, 46, 96–97, 113, 124; *see also* history
cultural misunderstanding 59, 64, 122

cultural relativism 95
cultural reproduction 48, 62, 96–98, 110–112, 119, 125
cultural selection 97–98
culture: to explain a 62, 68–69, 75, 79, 86
culture-gene coevolution 112–115, 119, 125

Darwin, Charles 10, 13, 69, 71, 112
Darwinian 71, 82, 95, 97, 101, 106, 110, 115–116
Darwinism 97, 126
Dawkins, Richard 14, 34–35, 96, 104, 118–119
Deacon, Terence 35, 63
defectors 105–106, 108
Dennett, Daniel 16, 35, 115–116, 119
diet 47, 105, 113
DNA 10–18, 56, 59, 91, 101, 112, 115–118
domain (of a cognitive module) 28–29
Dunbar, Robin 35, 56
Durham, William H. 120

emergence 9, 20, 28, 38, 40, 55, 63, 88, 121
emergent property 5, 54–57, 85, 91
encephalisation 22
environment 7–8, 12–13, 15–20, 22–24, 27–29, 31, 33, 37–38, 40–42, 44–47, 49, 52, 57, 68, 71, 74, 81–83, 91, 95, 110, 112–115, 117, 122, 124–125; of evolutionary adaptedness (EEA) 18, 47, 71, 105, 113
epidemiological model 48, 96, 112
ethnography 64–66, 68, 87, 92, 119
eusocial insects 103
Evans-Pritchard, Edward 92
evolution 5, 10, 13, 15–16, 20, *21*, 27, 34–35, 46, 49, 56, *70*, 71, 73, 82, 89, 95–97, 99–102, 104, 106, 108, 110–116, 118–119, 122, 124; cultural 5, 62, 70, 72, 75, 79, 92, 94–101, 106, 109–112, 114–116, 118–119, 124–125; human 1, 6, 9, 20, *21*, 22–23, 26, 28, 33–35, 37, 39, 63, 77, 99, 108, 114–115, 121
evolutionism 69, 71, 73–74, 75, 77, 95; cultural 79, 92, 97, 99, 101, 109, 111; social 68–75, 77, 94–96, 123
exaptation 49
exchange 12, 87–90

false belief task 32
Feldman, Marc 97, 119
fitness 28, 33, 42–43, 45, 48, 83–84, 99, 102–104; inclusive 104, 107

free-rider 44, 107–108
functionalism 68, 72–73, 77, 81–84, 86–87, 91–93, 95–96, 100, 109, 117, 123–124; *see also* structural-functionalism

Geertz, Clifford 62
gender 53, 65
group selection 101–104, 106–108, 110, 112, 114–115, 119, 125

Hamilton, William 103–104, 119
hard-wired 52, 99
Henrich, Joseph 63, 119
Hindu caste system 44–45
historicism 68, 72–73, 76, 80, 82–84, 92, 94–96, 123–124
history 1, 3, 6–10, 14, 20, 26, 39, 41, 45–46, 52, 60, 68–73, 75–77, 79–80, 86–88, 90–92, 95–97, 105, 123, 125; of life 6, 9–10, 12–13, 20, 40
hominids 24–25, 28
Homo erectus *21*, 25–28, 33
Homo habilis *21*, 24–25, 27
Hrdy, Sarah 35
human behaviour 1, 3–5, 8, 53, 55–56, 60, 66–67, 75–76, 79, 85, 87–88, 114–115
human nature 67, 79
humanities 2–4
hunter-gatherers 96, 105, 107, 111

imitation 52–53, 58, 90, 99
incest 88–89, 116–117
information 14–15, 17–20, 27, 29–30, 38, 40, 45–46, 51–53, 56–57, 63, 65–68, 73–74, 76–79, 87, 90, 122
innate 8, 16, 19, 30, 40, 45, 49, 72, 99–100, 114–115, 117
instincts 17, 19, 34, 44–45, 47, 51–52, 55–57, 60–61, 88–89, 105, 117
intelligence 6–7, 22, 118
intelligent design 16, 116–117
intentions 26, 31–34, 38, 44, 51–52, 54–55, 57–59, 61, 63, 85–86, 89, 121, 123
interpretation 34, 62, 73, 92
intuituitive knowledge 47, 69, 95, 99–100, 106

James, William 34
junk DNA 13–15

kin selection 103–104, 119
knowledge 4–5, 15–16, 18–21, 28–29, 31, 37, 38–42, 44, 46–48, 52, 53, 55, 62, 65–70, 72, 74–76, 80, 85, 92, 105,

110, 121–123; cultural 5, 15, 20–22, 26, 28, 34–35, 38–49, 52–54, 58–65, 67–68, 70–75, 77, 80–81, 96, 98–99, 108–111, 114, 121–124; genetic 15–20, 29–31, 34, 38, 40–44, 47–48, 55–61, 68, 85, 122; individual 15, 18–20, 22, 37–42, 44, 48, 54–57, 59–61, 65, 85, 98, 122, 124

lactose absorption 3, 113–114, 119
Lamarckian inheritance 98, 110, 114
language 5, 25, 27–28, 30–32, 35, 48–52, 59, 63–64, 76, 78, 87, 91, 116
Language Acquisition Device (LAD) 30–31, 49–50
learning 8–9, 19, 30–31, 38, 40, 64–65, 110, 124; cultural 28, 38–39, 62, 52, 85; social 38–39, 62
Lévi-Strauss, Claude 84–90, 92, 125
LUCA (Last Universal Common Ancestor) 9, *11*

McCauley, Robert 29
magic *see* belief in magic
making sense 4, 58, 61, 74
maladaptation 12, 25, 28, 42–44, 46–47, 83, 94–95, 97–101, 106, 108–109, 113–114, 117, 124–125
Malinowski, Bronislaw 77–81, 83, 85, 90–92
manufactured objects 26, 31, 38, 40, 58
marriage 87, 89–90
meaning 5, 8, 48, 50, 52–53, 55, 57, 61–62, 65–66, 84–87, 89–91, 116–117, 123, 125
meme 96, 108–112, 114–116, 119
memeticist theory of culture 96, 110–112, 115, 119, 125
memory 20, 27, 42, 57
mental representations 38, 54, 67, 124
mental states 26, 31–34, 50–52
mental time travelling 25–26
mentalisation 26, 32–33, 121; *see also* theory of mind
Mesoudi, Alex 98, 119
mind 8, 10, 16, 26, 31–32, 35, 38–40, 43–47, 51–60, 63, 65–67, 69, 76, 78–80, 84–85, 87–91, 96, 106, 109–111, 115–116, 121–122, 125; theory of (ToM) 26, 28, 30–35, 38–40, 49, 51–52, 54–55, 57, 59, 121
mirror neurons 31
modularity 26, 35
mutation 12–20, 23–24, 27–28, 39, 41–43, 82–83, 91, 97–98, 108, 111, 113, 115–118, 122

natural selection 5, 12–13, 15–16, 18, 27–29, 39, 41, 44, 46–47, 49, 67, 71, 82, 88, 95–98, 101, 105–108, 112–113, 115–116, 122, 124–125
Neolithic revolution 88, 113
nervous system 17, 19–20, 46, 52, 55–57, 65
neurons 22, 31, 54–57; *see also* mirror neurons
neurosciences 4, 35
niche construction 37–38
non-adaptation 12–13, 24, 42–44, 48; *see also* maladaptation

parasitic 43, 83, 108–110, 124
participant observation 65, 68, 78, 91
Pavlov, Ivan 57
Peirce, Charles S. 50
perception 32, 74, 76
philosophy 2, 16, 29, 34, 35, 38, 59, 63, 69, 92, 115, 119
Pinker, Steven 35
plasticity 18, 28
pointing 32, 52, 100
primate 3, 7, *11*, 17, *21*, 22–24, 27–28, 32, 34, 46, 56, 58
primitive 7, 65, 69–71, 73–74, 79, 95, 123
prisoner's dilemma 104–105
psychobiological needs 80–83
psychology 1, 4, 31, 66, 100, 114, 119

Quinn, Naomi 53, 63, 91

race 7–8
Radcliffe-Brown, Alfred R. 81, 92
religion 45, 68, 109
reproductive cycles 17–18, 20, 41, 118, 122
Richerson, Peter 62, 97, 119
Rizzolatti, Giacomo 31

science 1–4, 8, 35, 67, 71, 92, 95; social 1, 3–5, 35, 71–72, 92, 109, 115, 119
Scott-Philips, Thom 51, 63
selective advantage 33, 39, 50, 103, 110, 117
selfish gene 44, 104, 107, 112, 114
sexual reproduction 12–14, 111
social knowledge *see* social learning
social learning 38–39, 62, 69
social structure 81–82
spandrel *see* exaptation
Sperber, Dan 63, 119, 125
Stocking, George W. 91–92

Strauss, Claudia 53, 63
structural-functionalism 81–82, 84, 92
structuralism 68, 72, 84–85, 90–92, 94–96, 116, 123
sweet tooth 45, 47
symbiotic 43, 83, 103, 108–109, 124–125
symbol 8, 50–51, 57, 63, 65–66

tools: production and sharing of 24–26, 34, 39–40

tradition 2, 44, 58–59, 61–62, 68–69, 75–84, 92, 94, 101, 119, 123–124

unconscious 30, 40, 53, 57–58, 65, 67, 87, 90, 117

virus 39, 43, 46, 49, 53, 58, 83–84, 96

Westermarck, Edvard 88
Wittgenstein, Ludwig 59